AF541567

Soil Organic Matter

NIPA® GENX ELECTRONIC RESOURCES & SOLUTIONS P. LTD.
New Delhi-110 034

About the Author

Dr. A. Raja Rajan graduated from Tamil Nadu Agricultural University, (TNAU) Coimbatore and holds a doctorate in Soil Science & Agricultural Chemistry. He had his post-doctoral training on Soil Organic Matter at University of Florida, Gainesville, USA as an FAO Fellow. He has served in various positions in TNAU and was Dean at Agricultural College & Research Institute, Killikulam (TN), Dean, Faculty of Agriculture, TNAU, Coimbatore and Dean, Faculty of Agricultural Education and Research of Ramakrishna Mission Vivekananda Educational and Research Institute at its Coimbatore Campus. He has vast teaching and research experience in Agriculture, spanning over 40 years and has guided several post-graduate and doctoral students.

Soil Organic Matter
Key to Sustainable Soils

A. Raja Rajan
Formerly Dean
Faculty of Agriculture
Tamil Nadu Agricultural University
Coimbatore- 641003, Tamil Nadu, India

NIPA® GENX ELECTRONIC RESOURCES & SOLUTIONS P. LTD.
New Delhi-110 034

**NIPA® GENX ELECTRONIC
RESOURCES & SOLUTIONS P. LTD.**

101,103, Vikas Surya Plaza, CU Block
L.S.C.Market, Pitam Pura, New Delhi-110 034
Ph : +91 11 27341616, 27341717, 27341718
E-mail:newindiapublishingagency@gmail.com
www: www.nipabooks.com
For customer assistance, please contact
Phone: + 91-11-27 34 17 17 Fax: + 91-11- 27 34 16 16
E-Mail: feedbacks@nipabooks.com

ISBN: 978-93-95319-28-7

Composed and Designed by NIPA®.

Dedicated to
my fondest hopes:
Rida and Vishvesh

Preface

Soil organic matter is any material produced originally by living plants or animals that is returned to the soil and undergoes the decomposition process. Aside from providing nutrients and habitat to a multitude of organisms living in the soil, organic matter influences the physical and chemical environment of the soil and its overall health and quality.

Nutrient exchanges among organic matter, water and soil are essential to soil fertility and sustainable crop production. Where the soil is exploited for crop production without restoring the organic matter, soil fertility declines and the balance in the agro-ecosystem is wrecked. Thus, the real challenge will be to recognize management practices that nurture soil organic matter and ensure greater productivity and profitability to the farmers.

This book dwells on soil organic matter in its entirety: the composition, distribution, pools and reactive functional groups of soil organic matter; its decomposition, nutrient transformations and biochemistry of humus formation; its role in pedogenic processes; adsorption of organic compounds by clay; clay-organic matter complexes; humus - trace metals and humus - pesticides interactions; environmental significance of humic substances and characterization of soil organic matter. The potentials of nuclear techniques in the study of soil organic matter have been elucidated for the benefit of research scholars. Various management practices for building organic matter in soils have also been discussed. A compilation of qualitative and quantitative analytical procedures on organic matter complements the book.

It is believed that this book will be a useful source material for researchers, scholars and all stakeholders concerned with soil organic matter and sustainable agriculture.

I express my sincere appreciation of the wonderful support extended by my family during the writing of this book and to NIPA Genx Electronic Resources & Solutions (P) Ltd., for bringing out this book elegantly.

Jan, 2023 **A. Raja Rajan**

Contents

1

Introduction

The term soil organic matter (SOM) includes any material that is produced originally by soil organisms (plant or animal) that is returned to the soil and undergoes decomposition process. At any point in time, it consists of an array of materials, ranging from the intact original tissues of plants and animals to the substantially decomposed mixture of materials known as *humus* (Fig.1.1). Most soil organic matter originates from plant tissues. Plant residues contain 60 - 90 per cent of moisture and the remaining dry matter consists mostly of carbon, oxygen, hydrogen, with smaller amounts of sulphur, nitrogen, phosphorus, potassium, calcium and magnesium. Though present in small amounts, these nutrients are of great significance from the standpoint of soil fertility.

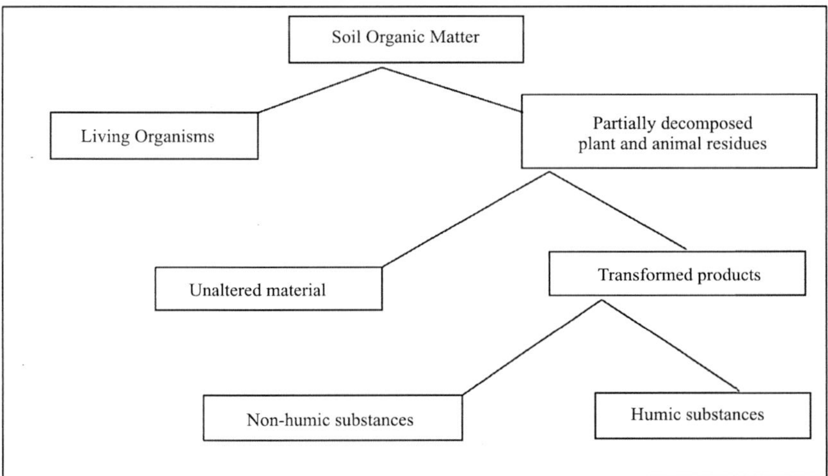

Fig. 1.1: Components of soil organic matter

Soil organic matter is composed of a number of components. These components include, in varying proportions and several intermediate stages, an active organic fraction constituting microorganisms (10 - 40 per cent) and resistant or stable organic matter (40 - 60 per cent), also referred to as *humus*.

In practical terms, organic matter may be divided into above-ground and below-ground fractions. While the above-ground organic matter comprises plant and animal residues, below-ground organic matter comprises living soil fauna and microflora, partially decomposed plant and animal residues and humic substances. The C:N ratio indicates the type of material and ease of decomposition; hard, woody materials have a high C:N ratio and are more resilient than soft leafy materials with a low C:N ratio. In spite of the fact that soil organic matter can be readily fractionated into its constituent fractions, these fractions do not represent static end products. Instead, the amounts reflect a dynamic equilibrium. The total amount and fractions of organic matter in the soil are influenced by soil properties and by the quantity of annual inputs of plant and animal residues to the soil ecosystem. Soil properties such as texture, pH, temperature, moisture, aeration, clay mineralogy and soil biological activities all determine the rate of decomposition and accumulation of soil organic matter in a given soil ecosystem. The irony is that soil organic matter influences many of these very same soil properties!

Organic matter present on the soil surface as raw plant residues protects the soil from the effects of rainfall, wind and Sun. By removal or burning of residues, the soil is exposed to the vagaries of climate. By removal or burning of residues, the soil organisms are deprived of their prime energy source. Soil organic matter serves several functions. From the practical agriculture point of view, soil organic matter serves: (i) as a dynamic nutrient buffer and (ii) as a medium to improve soil structure, maintain tilth and minimize erosion.

Because soil organic matter is derived mainly from plant residues, it contains all essential plant nutrients and, therefore, is a storehouse of plant nutrients. The stable organic fraction (humus) binds and holds nutrients in a plant-available form and, upon decomposition, organic matter releases these plant-available nutrients. For this nutrient cycling system to sustain, the rate of addition of organic materials from crop residues, manures and other sources must equal the rate of decomposition of these materials and must take into account the rate of uptake by plants and the rate of leaching and erosion losses.

When the rate of addition is lesser than the rate of decomposition, soil organic matter declines. On the other hand, when the rate of addition is higher than the rate of decomposition, soil organic matter increases. The term 'steady state' denotes a condition wherein the rate of organic matter addition is equal to the rate of organic matter decomposition.

When it comes to improving soil structure, the active soil organic components, along with some of the resistant components and microorganisms (especially, fungi), are credited with binding soil particles into larger aggregates.

Aggregation is crucial for good soil structure, aeration, water infiltration and resistance to erosion and crusting.

Soil aggregation has always been linked with soil carbon levels. In the recent past, techniques have been developed to fractionate carbon on the basis of lability (ease of oxidation), with the realization that these sub pools of carbon, besides influencing soil physical stability, may be much better indicators of carbon dynamics than the total carbon values in agricultural production systems. The labile carbon fraction has been proved to be an indicator of key soil chemical and physical properties. This fraction has been shown, after simulated rain in laboratory experiments, to primarily control aggregate breakdown in red clays (Ferrosols) as per the percentage of aggregates measured in the surface crust. The stable fraction of soil organic matter contributes primarily to cation exchange capacity (by holding nutrients) and soil colour. Since the decomposition of this fraction of soil organic matter is very slow, its influence on soil fertility is less as compared to the active organic fraction.

Soil organic matter consists of three primary constituents including fresh plant residues and small living soil organisms, decomposing active organic matter and stable organic matter (humus). It is a reservoir of nutrients for crop plants. It facilitates soil aggregation, promotes exchange of nutrients, enhances water infiltration and retention of moisture in soil and reduces soil compaction and surface crusting. The components vary in relative proportions and have several intermediate stages (Fig. 1.2).

Plant residues such as leaves, manure or crop residue littered on the soil surface are not *per se* considered as soil organic matter and are usually removed from soil samples before analysis, by sieving through a 2 mm wire mesh. It is possible to estimate soil organic matter content in the field and determine in the laboratory the amounts of nitrogen, phosphorus and sulphur mineralized that are available for crop plants and therefore fine-tune fertilizer recommendations.

Soil organic matter influences the rate of surface-applied herbicides as well as soil pH necessary to effectively control weeds. It also influences the magnitude of herbicide carryover for future crops and amount of lime necessary to raise soil pH.

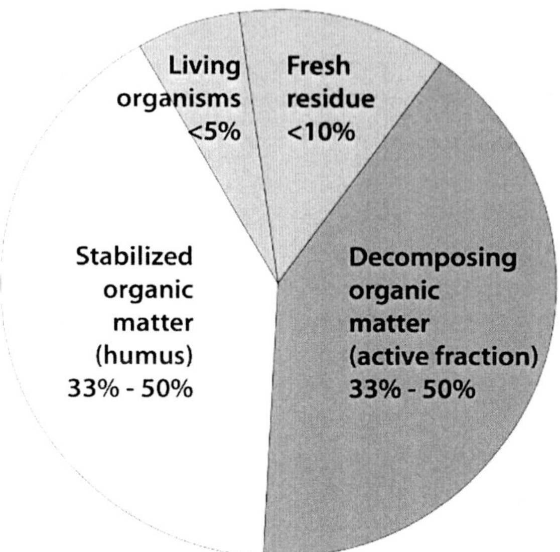

Fig. 1.2: Relative proportions of soil organic matter components

1.1. Factors Affecting Soil Organic Matter

Rainfall, temperature, moisture and soil oxygen levels (aeration) all affect the decomposition rate of soil organic matter. Factors such as climate and soil texture that affect soil organic matter cannot be manipulated. Organic matter decomposes quickly in warm and humid climates and slowly in cool, dry climates. The decomposition is fast when the soil is well aerated and is much slow in saturated soils.

In comparison with the soils formed under forests, soils under grasslands (prairie vegetation) typically have at least twice as much organic matter levels, because organic material is added to topsoil not only from top growth but also from roots that die back every year. Forest soils usually have low organic matter levels owing to two main factors:

1. The per acre root mass production by trees is insignificant in comparison with grasses.
2. Unlike grasses, trees do not die back and decompose every year. On the contrary, much of the organic material in a forest is tethered in the tree's wood instead of being returned to the soil every year.

1.2. Soil Organic Matter Management

Where biomass production is higher and where organic material additions occur, soil organic matter content is generally bound to increase. Plant residues having a high nitrogen content (low C/N ratio) decompose more quickly than those with a low nitrogen content (high C/N ratio) and therefore the soil organic matter levels do not increase as quickly. When the soil is tilled excessively, soil aggregates are destroyed and the rate of decomposition of soil organic matter is increased. Stable soil aggregates foster active organic matter and safeguard stable organic matter from the vicissitudes of rapid microbial decomposition. Measures that increase soil moisture, soil temperature and optimal aeration accelerate soil organic matter decomposition.

Management practices adopted on field can lead to either degradation or increase in soil organic matter. Some key management measures that can increase soil organic matter are listed below. A more detailed discussion is given in Chapter 19.

- Adopting no-till cropping systems with cover crops, solid organic materials and assorted rotations with high residue crops and perennial grasses and legumes.
- Reducing tillage that causes a flurry of microbial activity leading to speeding up of organic matter decomposition.
- Reducing erosion by suitable measures. Since most soil organic matter is associated with the topsoil, when soil is eroded, organic matter also is lost along with it. Erosion control measures save soil and soil organic matter in tandem.
- Testing soil and fertilizing accordingly. Proper fertilization ensures increased root and top growth of crop plants. Even when the top growth is removed, increased root growth can help build - or at least maintain - soil organic matter.
- Including perennial grasses in the production system provides for annual die back and re-growth of perennial grasses and their vast, fibrous root systems contribute organic matter to soil and serve as a binding agent in soil aggregation.

1.3. Soil Organic Matter Relationship to Soil Function

Depending on site-specific management and climatic factors, mineralization rates and loss of soil organic matter can increase significantly under favorable conditions of aeration, temperature and moisture. The key soil functions that soil organic matter is associated with include:

- **Nutrient Supply:** During the decomposition of organic matter, plant nutrients are released to crop plants in readily available forms. In a medium textured soil (silt and loams), every per cent of soil organic matter in the top 6 inches of soil releases around 10-20 kg of nitrogen, 1 to 2 kg of phosphorus and 0.4 to 0.8 kg of sulphur, every hectare, every year.

- **Water-holding Capacity:** Fairly akin to a sponge, organic matter has the ability to absorb and hold water even up to 90 per cent of its weight. The greatest advantage of this property is that it releases almost all the water it holds for plants to use. In sharp contrast, though clay holds significant quantities of water, much of it is unavailable to crop plants.
- **Soil Aggregation:** Organic matter improves soil structure through promoting soil aggregation. With better soil structure, water infiltration through the soil will be better and thus the soil's ability to hold water will also improve.
- **Erosion Prevention:** With increased organic matter erosion is reduced, owing to increased water infiltration and stable soil aggregates.

1.4. Organic Material Required to Increase Soil Organic Matter

It is common knowledge that if the rate of organic material addition is less than the rate of decomposition, soil organic matter levels will go down. On the contrary, when the rate of organic material addition is greater than the rate of decomposition, soil organic matter levels will rise up. When the rate of organic matter addition from organic materials is on par with the rate of decomposition, it is referred to as **steady state**.

One hectare of soil, 6 inches deep, weighs approximately 2.24 x 10^6 kg. In other words, 1% soil organic matter weighs about 22,400 kg per ha. Under normal conditions it takes at least 10 kg of organic material to decompose into 1 kg of organic matter. Thus, it takes at least 224 tonnes of organic materials applied (or returned) to the soil to add about 1% of stable organic matter in favorable conditions.

2

Soil Organic Matter Pools

Soil organic matter makes up only a few per cent of most soils, but it has a great deal of influence on soil properties and on agricultural productivity.

2.1. Quantity of Organic Matter

Based on the content of organic matter, soils are identified as mineral or organic. Mineral soils dominate most of the world's cultivated land and may contain anything from a trace to 30 per cent of organic matter. Organic soils are naturally rich in organic matter, chiefly due to climatic reasons. Although they contain more than 30 per cent organic matter, it is precisely for this reason that they are not vital cropping soils.

The level of soil organic matter in a particular soil is mainly due to natural factors such as temperature (cool places accumulate more organic matter), soil texture (clayey and silty soils tend to have more organic matter than sandy ones), and the drainage conditions (poor drainage favours soil organic matter build up). However, management practices like frequent tillage operations, long periods of barren land and removal of crop residues without replenishment all contribute to abatement of soil organic matter.

2.2. Composition of Organic Matter

Soil organic matter is composed of plant and animal residues in different stages of decomposition, cells of soil microorganisms and substances that are so well-decomposed that it is impossible to recognize what they were originally at the start. Living organisms are also considered to be a vital cog in soil organic matter because of their huge role in contributing organic residues to the soil and in the formation of more stable fractions of organic matter. Besides plant roots, an array of soil animals (rodents, earthworms, mites, etc.) furnish organic materials to the soil that eventually partake in the soil organic matter cycle. All the four main processes in the organic matter cycle depend on soil microbes: decomposition of organic residues, release of nutrients (mineralization), release of carbon dioxide (respiration) and transfer of carbon among different soil organic matter 'pools'.

2.3. Kinds of Organic Matter

In addition to organic matter that is alive, there are three types, or pools, of "dead" soil organic matter: active, slow, and passive, or, stable (Fig. 2.1). These pools have different histories, characteristics and turnover times. These are adjudged by the time it takes for their complete decomposition.

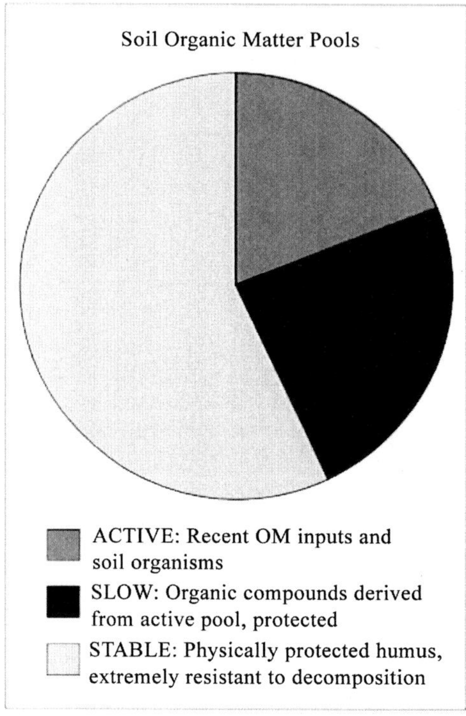

Fig. 2.1: Typical distribution of SOM pools
(Actual proportions may vary with soil type and management practices)

Active soil organic matter is primarily made up of fresh plant and animal residues that are in the early stages of decomposition and of soil organisms. They break down in a very short period of time, varying from a few weeks to a few years. The active organic matter is involved in a lot of biological activity, helping to develop a soil's slow organic matter pool, and is very important for nutrient release.

Passive or stable soil organic matter, also known as *humus*, is not biologically active; therefore, it provides very little food for soil organisms. It may take hundreds - or even thousands - of years before it is fully decomposed!

Slow soil organic matter is intermediate between active and passive soil pools, consisting primarily of detritus, partially broken-down cells and tissues that can only gradually decompose. It is rather resistant to decay and may take a time lag of few years to a few decades for complete breakdown. This slow turnover pool of organic matter greatly influences both the physical condition and nutrient buffering capacity of a soil. Starkly differing from the stable soil organic matter pool, the slow pool is a vital source of nutrients, particularly nitrogen, phosphorous and sulphur.

The passive fraction of soil organic matter is humus. Humus is a dark, complex mixture of organic substances that have been significantly modified from their original form over a period of time and also contains other substances that have been synthesized by soil organisms. Usually, humus constitutes a majority of the total soil organic matter and is relatively stable with the passage of time. The ability of a soil to retain nutrients and water is credited to the humus. Humus can increase the availability of trace elements to plants by supplying organic substances to the soil solution that can serve as chelates of trace elements.

2.4. Active Soil Organic Matter

Active soil organic matter is the readily digestible and easily decomposed portion of fresh organic residues and is 'fresh meat' to microbes. As it is decomposed by soil organisms, it helps stabilize soil aggregates, releases nutrients by mineralization and provides food for microbial activity, leading to enhanced plant growth.

The amount of active organic matter in the soil can change quickly, in just a year or two, and is highly influenced by soil management practices. To sustain the same level of active soil organic matter, a constant supply of fresh organic materials is needed, usually from growing plants. Roots, residues and cover crops contribute to active organic matter. Soil must also be managed to minimize the loss of organic matter through oxidation (from aggressive tillage) and erosion (from ground being left bare). The amount of active soil organic matter and its proportion to total soil organic matter are good indicators of soil health.

2.5. Functional Soil Organic Matter Pools

Soil organic matter is the total organic fraction of a soil. It is composed of living microbial biomass and a very complex, heterogeneous mixture of organic compounds and minerals, including residues of plant and microbial origin at different stages of decomposition and various degrees of stabilisation. It is heterogeneously distributed between the soil compartments and varies very widely in quality, availability and stability. The properties of its constituent compounds are highly variable, ranging from readily-degradable compounds to compounds that are physically resistant to decomposition. Current conceptual models differentiate functional soil organic matter pools in *active*, *intermediate* and *passive* pools on the basis of their different turnover rates, or equivalently, their residence times.

In order to understand the dynamics of soil organic matter, it is imperative to identify, isolate and characterize functional soil organic matter pools. This can be accomplished using methods, such as physical fractionation. Physical fractionation techniques include size and density-based separation of organo-mineral complexes. By using complementary methods for fractionation on the basis of particle size and specific density, it is possible to separate *uncomplexed*, *primary* and *secondary complexed* soil organic matter (Fig. 2.2).

Uncomplexed soil organic matter is not associated with minerals and is separated by partitioning in heavy liquids (as polytungstates, iodates or bromoform) with densities of 1.6 –2.0 g cm^{-3}. The uncomplexed organic matter in soils consists of both *free* and *occluded* soil organic matter. Free soil organic matter includes loose organic particles, while occluded soil organic matter is physically protected by aggregates. By complete dispersal of the soil (e.g., by sonication in water), complexed soil organic matter is obtained. Since the highest concentrations of soil organic matter (50 - 75% C) are associated with clay-sized particles, these are of particular importance when studying the accumulation of C in soils. Secondary complexed soil organic matter consists of micro- (< 250 μm) and macro-aggregates (> 250 μm).

Functional soil organic matter pools (uncomplexed and complexed soil organic matter) are stabilised in soils by specific mechanisms with certain turnover rates. The stabilisation of soil organic matter is what is referred to as the "*protection of organic matter from mineralisation*". Three general mechanisms of soil organic matter stabilisation have been described: (1) *Recalcitrance* (2) *Spatial inaccessibility* (3) *Interaction.*

Recalcitrance is the selective preservation of soil organic matter due to its structural composition / molecular-level characteristics. Recalcitrance is primarily associated with plant litter and rhizodeposits, microbial products, humic polymers and charred soil organic matter, which are all relatively slowly degraded by microbes and enzymes. However, "*recalcitrance*" implies only a relative stability - the microbial community in soil will eventually degrade any natural soil organic matter.

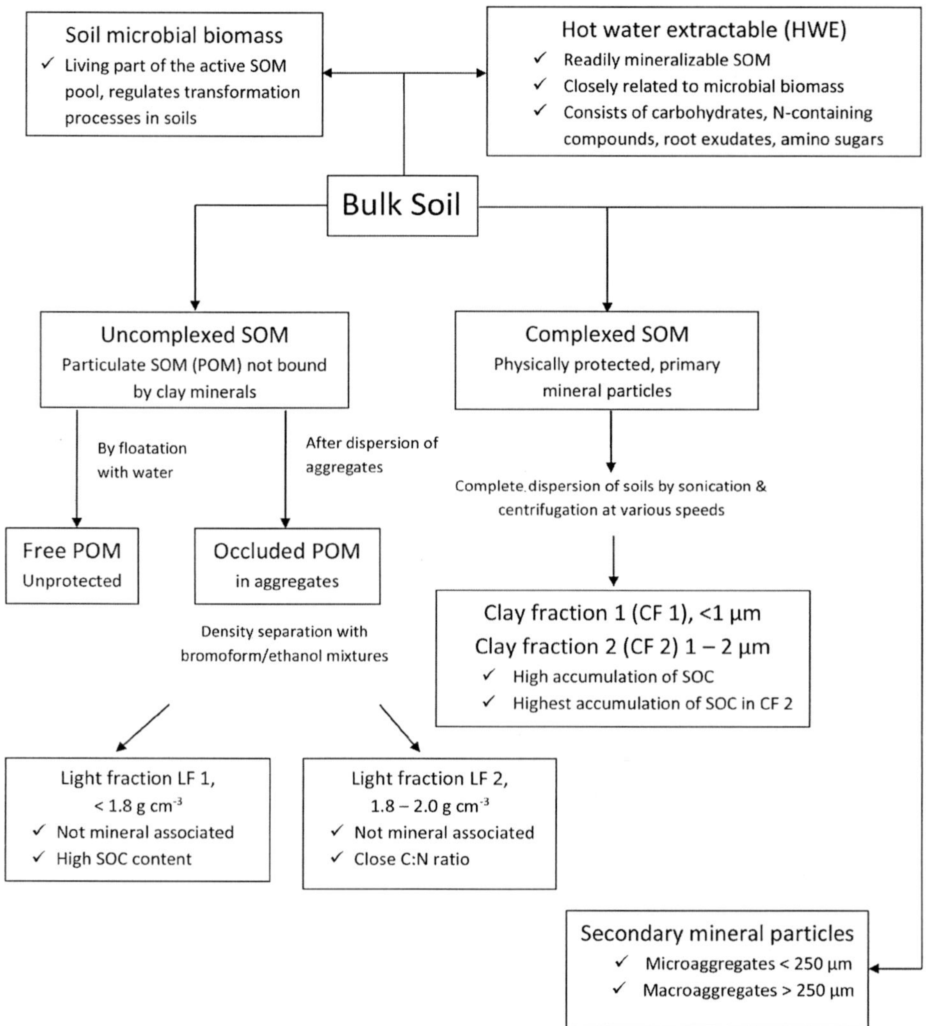

Fig. 2.2: Conceptual model of functional soil organic matter pools

Spatial inaccessibility refers to various processes that cause the physical occlusion of soil organic matter, rendering it inaccessible to microbes and degradative enzymes. These include interactions with aggregates, effects related to hydrophobicity and encapsulation within organic macromolecules. Humic substances have a tendency to form organo-mineral complexes. Metabolic binding agents produced by the soil biota can play a vital role in the formation of larger aggregates.

Interactions between the soil organic matter and minerals and metal ions reduce the availability of soil organic matter to microorganisms and enzymes.

Clay minerals and / or amorphous hydroxides of Fe and Al play a major role in these processes because of their reactive hydroxyl groups. In soils, where the interaction between soil organic matter and minerals or metal ions is significant, stabilised soil organic matter pools with relatively high mean ages and low turnover rates are observed.

3

Composition, Distribution and Functions of Soil Organic Matter

Soils vary greatly in their organic matter content. For all practical purposes the organic matter content of the soil parallels the soil N content. The C/N ratio of soil organic matter generally falls in the range of 10 to 12. Because of the ease with which Kjeldahl determination can be made, N is often used as an index of soil organic matter content.

3.1. Composition of Organic Matter in Soil

Organic matter constitutes 1 to 6% of the topsoil weight of most upland soils. In a typical grassland soil, 5 to 6% of organic matter may be present in the top 15 cm. Top soils with less than 1% organic matter are mostly limited to desert regions. On the contrary, the organic matter content of soils in low, wet areas may be as high as 90% or even more. Soils with more than 12 - 18% organic carbon (approximately 20 - 30% organic matter) are called *organic soils*.

3.2. Distribution of Organic Matter in Soil

Natural processes leading to the development of soils having variable organic matter contents are related to the factors of soil formation:

$$OM = f\,\{\text{time, climate, vegetation, parent material, topography} \ldots\}$$

where *f* stands for "function of" and the dots indicate that there may be other factors involved. Hans Jenny[1], a Swiss-born Soil Scientist and expert on Pedology treated each soil forming factor as an independent variable. For e.g., $OM = f\,\{\text{Climate}\}_{\text{time, vegetation, parent material, topography} \ldots\ldots}$ According to Jenny, the order of importance of soil forming factors in determining the organic matter and N contents of loamy soils was: climate > vegetation > topography > parent material > time.

---

[1]Hans Jenny. 1994. Factors of Soil Formation: A System of Quantitative Pedology. Dover Publications Inc., New York.

3.2.1. Time

Organic matter does not accumulate indefinitely in well-drained soils and with time an equilibrium level is attained that is governed by the soil forming factors of climate, topography, vegetation and parent material. The numerous combinations under which the various factors operate account for the great variability in the organic matter content of soils.

The rate of organic matter accumulation is rapid during the first few years, diminishes slowly and reaches equilibrium in periods of time which vary from as little as 110 years for the fine textured parent material to as much as 1500 years for sandy areas. The explanations given for the establishment of equilibrium levels of organic matter in soil are: (1) Organic colloids [e.g., humic acids] are produced which resist attack by microbes, (2) humus is protected from decay through its interaction with mineral matter [e.g., polyvalent cations and clay] and (3) a limitation of one or more essential nutrients [N, P, S] places a cap on the amount of stable humus that can be synthesized.

The variation in the quantity of soil organic matter with time can be described by the simple equation proposed by Jenny (1941): $\delta C/\delta t = A - kC$, where C is the organic matter pool at time t, A is the mass of organic matter added and k is the coefficient of decomposition of organic matter. A valid estimate of soil organic matter pool, therefore, requires a good measurement of primary production and, above all, of the input of debris to soil (A). The size of the soil organic matter pool is generally correlated positively with rainfall and temperature, but more specifically to the species composition of the plant cover, as well as to the biological type of the locally dominant species controlling the chemical composition (e.g., lignin and nitrogen concentrations) of dead organic matter which is the food for microorganisms. Moreover, the spatial structure of the vegetation at different scales has a strong impact on the horizontal distribution of soil organic matter and nutrients, and consequently, on the ecosystem functioning and dynamics.

3.2.2. Climate

Climate is the most significant single factor that determines the array of plant species at any given location, the quantity of plant material produced and the intensity of microbial activity in the soil. Consequently, this factor plays a prominent role in determining the organic matter levels. A humid climate leads to forest associations and the development of Spodosols and Alfisols; a semi-arid climate leads to grassland associations and development of Mollisols.

Studies on the effect of climate on organic matter and nitrogen levels in soil have shown that the nitrogen content of soils decreased 2 to 3 times for every

10 °C rise in mean annual temperature. The effect of increasing rainfall on soil organic matter content is to promote greater plant growth, and consequently, production of larger quantities of raw material for humus synthesis.

3.2.3. Vegetation

Soils formed under grass vegetation (prairie) usually have organic matter levels at least twice as much as those under forests because organic materials are added to the top soil from both the top growth and from the roots that die back every year. Soils formed under forests usually have low organic matter levels due to trees producing a much smaller root mass per acre than grass plants and trees do not die back annually and decompose every year.

Inputs of plant carbon in the form of shoot, roots and the associated mycorrhizal fungi are indispensable for soil organic carbon stock and turnover. Soil fertility can affect both the amounts and proportions of plant carbon inputs to the soil through these pathways. Field experiments have shown that soils with higher fertility sequestered more shoot-, root- and arbuscular mycorrhizal fungi (AMF)-derived carbon, which were mainly driven by greater soil microbial biomass. Irrespective of soil fertility, roots contributed the most (44%) to new soil organic carbon formation, while shoot (28%) and AMF (28%) exerted similar but lower contributions. The disproportionately large priming effects relative to new soil organic carbon accumulation induced by AMF led to net soil organic carbon losses, especially in soils with higher fertility. Overall, the plant carbon inputs through shoot, root, and mycorrhizal pathways have differential impacts on soil organic carbon turnover.

The soil microbiome community plays a vital role in processing and storing nitrogen and carbon and is affected by the above-ground vegetation through differences in chemistry, structure and mass of plant litter, root physiology and dominant mycorrhizal associations. Shotgun metagenomic sequencing studies used to quantify the abundance and distribution of gene families involved in soil microbial nitrogen and carbon cycling beneath three deciduous hardwood tree species, *viz*., ectomycorrhizal-associated *Quercus rubra* (red oak), ectomycorrhizal-associated *Castanea dentata* (American chestnut) and arbuscular mycorrhizal-associated *Prunus serotina* (black cherry) have proved that, of the three species, Chestnut had the most distinct soil microbiome associations, both in functional and taxonomic terms, with subdued functional genes in the nitrification, denitrification and nitrate-reduction pathways. These changes were attributed to low inorganic nitrogen availability in chestnut stands, as soil was modified by poor, low-nitrogen litter quality as compared to red oak and black cherry soils. This highlights the genetic framework underlying tree species' control over soil microbial communities and below-ground carbon and nitrogen metabolism and may enable land managers to select tree species to maximize carbon and nitrogen storage in soils.

3.2.4. Parent material

Parent material is effective mostly through its influence on texture. For any given climate zone, the organic matter content depends upon the textural properties, provided vegetation and topography are constant. Humic substances fixed in the form of organo-mineral complexes help preserve organic matter. The organic matter content in soils follows the order: heavy textured soils > loamy soils > sandy soils.

The type of clay mineral can affect the retention properties of soils. Montmorillonite clays with high adsorption capacities for organic molecules are particularly effective in protecting nitrogenous constituents against microbial attack.

3.2.5. Topography

Topography affects soil organic matter content through its influence on climate, run off, evaporation and transpiration. Local variations in topography, such as knolls, slopes and depressions, modify the plant microclimate. Soils formed in depressions, where the prevalent climate is "locally humid", are richer in organic matter content than the ones formed on the knolls where the climate is "locally arid". Poorly drained soils are usually high in organic matter, because the anaerobic conditions that prevail during wet periods of the year prevent destruction of organic matter.

3.2.6. Cropping

Though not always, organic matter levels usually decline when soils are first placed under cultivation. Improved aeration, resulting from cultivation, may lead to increased microbial activity and loss of organic matter. A temporary spurt in rate of respiration occurs every time an air-dried soil is wetted. Since significant amounts of fresh soil frequently go through wetting and drying cycles through cultivation, organic matter lost by this process could be palpable. A major effect of cultivation in stimulating microbial activity may be the exposure of organic matter previously not accessible to microbial attack.

3.3. Functions of Organic Matter in Soil

Organic matter contributes to plant growth through its effect on the physical, chemical and biological properties of soil (Fig. 3.1). The three-some functions of soil organic matter are not isolated and static, but dynamic interactions occur between the three major components. The properties of humus and associated effects on soil are given in Table 3.1.

3.3.1. Availability of Nutrients for Plant Growth

Organic matter has significant effect on the availability of nutrients for plant growth, both directly and indirectly. Besides serving as a source of nitrogen, phosphorus and sulphur through its mineralization by soil microbes, organic matter influences the supply of nutrients from other sources. For example, organic matter is required as an energy source for N-fixing bacteria; accordingly, the amount of N_2 fixed will be influenced by the quantity of energy available in the form of carbohydrates.

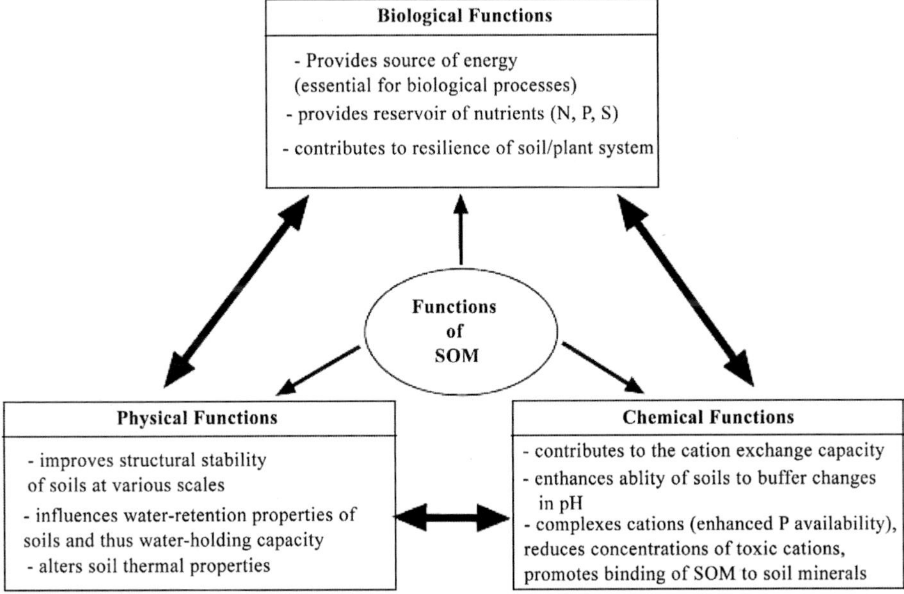

Fig. 3.1: Functions of soil organic matter with interactions between different soil functions

The availability of phosphorus in soil is often limited by fixation reactions which convert monophosphate ion ($H_2PO_3^-$) to insoluble forms (Ca-phosphates in calcareous soils and Fe- and Al-phosphates in acid soils). Adsorption of phosphorus by clay minerals can affect phosphate availability to crop plants in neutral as well as slightly acid conditions. Availability of soil phosphate is enhanced by additions of organic matter, presumably because of chelation of polyvalent cations by organic acids and other decay products. Naturally occurring chelating agents also influence the availability of micronutrients in much the same way as inorganic phosphates.

3.3.2. Effect on Soil Physical Condition

Humus has profound effect on the structure of many soils and when is lost, soils tend to become hard, compact and cloddy. Seedbed preparation and tillage operations are usually easier and effective when humus levels are adequate.

Table 3.1: General properties of humus and associated effects in the soil

Property	Remarks	Effect on soil
Colour	Typical dark colour of many soils is due to SOM	May facilitate warming
Water retention	SOM can hold water up to 20 times its weight	May significantly improve moisture-retention properties of sandy soils
Combination with clay minerals	Cements soil particles into structural units called aggregates	Permits exchange of gases; Stabilizes structure; increases permeability
Chelation	Forms stable complexes with Cu^{2+}, Mn^{2+}, Zn^{2+} and other polyvalent cations	May enhance availability of micronutrients to plants
Buffer action	Exhibits buffering in slightly acid, neutral and alkaline ranges	Helps maintain uniform reaction in soil
Cation exchange	Range for humus: 300 to 1,400 meq100 g^{-1}	May increase CEC of soil
Mineralization	Decomposition of SOM yields CO_2, NH_4^+, NO_4^{3-}, PO_4^{3-} & SO_4^{2-}	Source of nutrients for plant growth
Combination with organic molecules	Affects bioactivity, persistence and biodegradability of pesticides	Modifies application rate of pesticides for effective control

Humus favourably influences aeration, water holding capacity and permeability. With the addition of easily decomposable organic residues, complex organic compounds are synthesised (e.g., polysaccharides) and they bind soil particles into structural units called, 'aggregates' which help to maintain a loose, open, granular condition. Infiltration and percolation of water is then better. Large pores facilitate far better exchange of gases between soil and atmosphere. Plants roots get a continuous supply of O_2 for respiration and growth. The effect of organic matter on soil water retention is far greater in coarse textured soils than in fine textured ones.

Soil colour and organic matter content are positively correlated. Higher the soil organic matter content, darker is the soil colour and *vice versa*. Dark-coloured, high soil organic matter containing soils hold comparatively larger amounts of water than the light-coloured soils with low organic matter. The thermal conductivity of a soil is largely influenced by a combination of water content, texture and colour.

3.3.3. Effect on Soil Erosion

Humus enhances the ability of the soil to resist erosion. Firstly, it empowers the soil to hold more water. Secondly, by promoting soil granulation it helps to maintain large pores through which water can enter and percolate downward. In a granular soil, individual soil particles are not easily washed along by moving water.

3.3.4. Source of Energy

Both macro- and microfaunal organisms thrive on organic matter as their energy source. The population of bacteria, fungi and actinomycetes in the soil is related to humus content. Earthworms and other fauna are strongly affected by the quantity of plant residues returned to the soil. Earthworms, in particular, are important agents in producing good soil structure. By constructing extensive channels they serve not only to loosen the soil, but to improve aeration and drainage.

Higher plants are considered primary producers as they capture energy and CO_2. Debris from dead plants is degraded by the soil fauna and microflora, which are considered the primary, secondary and tertiary consumers (Fig. 3.2). They liberate energy and CO_2 and produce humus. About 80 - 90% of the total soil metabolism is carried out by the microflora.

One of the most fundamental functions of soil organic matter is provision of metabolic energy which drives soil biological processes. In crux, it is the transformation of carbon by plant, micro- and macro-biological processes which provides energy and establishment of a cycle that connects above- and below-ground energy transformations is the corollary.

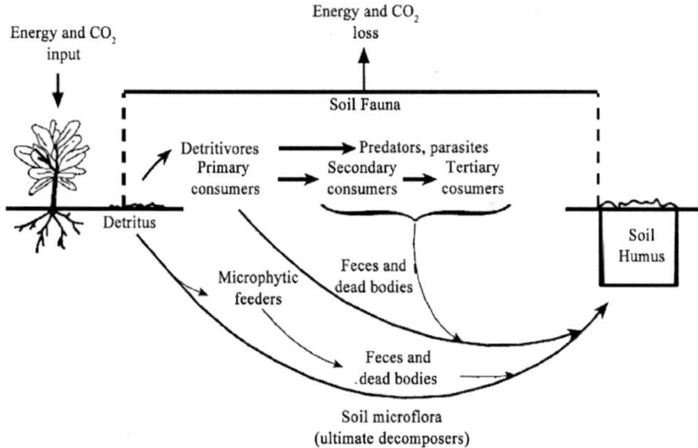

Fig. 3.2: General pathway for the breakdown of higher plant tissue

3.3.5. Buffering and Exchange Capacity

Colloidal humic substances account for 20 to 70% of the exchange capacity of many soils. Functional groups [e.g., carboxylic acids] of soil organic matter are main contributors to the cation exchange capacity as they provide negatively charged sites. Aside from functional groups, smaller particle size fractions (especially, the organo-mineral clay fractions) influence the cation exchange capacity much more than the coarser fractions. Charcoal has been shown to be a potentially important contributor to increasing cation exchange capacity.

There is generally a good correlation between buffering capacity and soil organic matter content. Functional groups are believed to be responsible for soil organic matter to act as a buffer over a wide range of pH values.

3.3.6. Adsorption of Pesticides and Other Organic Chemicals

Organic matter influences the behaviour of organic pesticides in soil, including effectiveness against target species, phytotoxicity to subsequent crops, leachability, volatility and biodegradability. The soil factor that is most intensely related to the sorption of most herbicides is organic matter content.

4

Soil Carbohydrates and Lipids

Carbon is the cornerstone of all life on Earth. Plant tissues and microbial cells contain approximately 40 to 50% of carbon on dry weight basis. Yet, the ultimate source is the CO_2 of the atmosphere. The CO_2 is converted into organic carbon by the action of photoautotrophic organisms (higher plants and green algae). It has been estimated that the vegetation on earth's surface consume some 1.3 x 10^{14} kg of CO_2 per annum which is about $^1/_{20}$ of the total CO_2 supply of the atmosphere or $^1/_{1000}$ of that in the ocean.

4.1. The Carbon Cycle

Soil carbon constitutes the largest terrestrial pool of carbon. Most carbon is stored on Earth in rocks and sediments. The rest of the carbon is in the ocean, atmosphere and in living organisms. Carbon cycles through these sinks.

The carbon cycle depicts the pathway in which carbon atoms continuously travel from the atmosphere to the Earth and then back into the atmosphere (Fig. 4.1). The Earth and its atmosphere form a closed environment and, therefore, the amount of carbon in this closed system does not change. Carbon is constantly in flux, irrespective of where it is located - in the atmosphere or on Earth. Carbon reaches back into the atmosphere when organisms die, or volcanoes erupt, fires blaze, fossil fuels are burnt, and through an array of other mechanisms.

Carbon is present in the atmosphere as carbon dioxide. Carbon enters the atmosphere through natural processes such as respiration and industrial processes such as burning fossil fuels. The CO_2 is absorbed by autotrophs such as green plants. The process of photosynthesis involves the absorption of CO_2 by plants to synthesize carbohydrates. The equation is as follows:

$$6CO_2 + 6H_2O + \text{Energy} \rightarrow (CH_2O)_6 + 6O_2\uparrow$$

Carbon compounds thus produced are passed along the food chain from the producers to consumers. When living organisms respire, the carbon exits their body in the form of CO_2. Animals consume plants and, thus, incorporate carbon into their system. Microbes use organic matter as their energy source and incorporate soil carbon into their body tissues.

When animals and plants die, their bodies that are constituents of soil organic matter are decomposed by soil microbes and carbon is released back into the atmosphere as CO_2. The decomposers eat the dead organisms and return the carbon from their body back into the atmosphere. The equation for this process is:

$$(CH_2O)_n + nO_2 \rightarrow nCO_2\uparrow + nH_2O$$

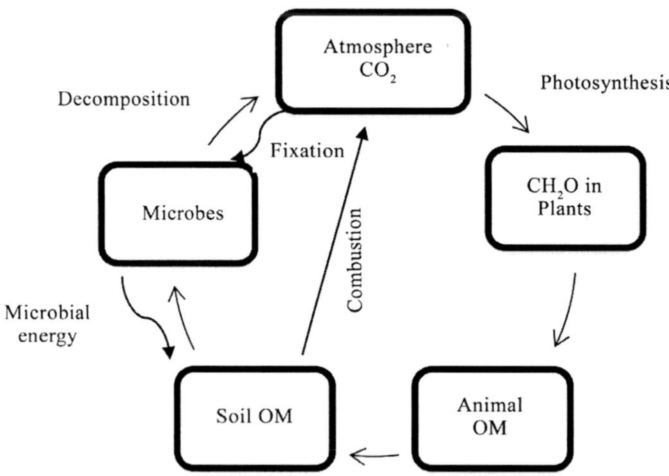

Fig. 4.1: The carbon cycle

4.2. Soil Carbohydrates

Carbohydrates constitute 5 to 25% of the organic matter in most soils. Plant residues furnish carbohydrates as simple sugars, hemicelluloses and cellulose which are all but decomposed by bacteria, actinomycetes and fungi, which in turn synthesize polysaccharides and other carbohydrates of their own.

4.2.1. Significance of Carbohydrates

The significance of carbohydrates in soil is primarily because of the ability of the complex polysaccharides to bind inorganic soil particles into stable aggregates. Carbohydrates also form complexes with metal ions and serve as building blocks for humus synthesis. A few of the sugars invigorate seed germination and root elongation. Cation exchange capacity (attributed to COOH groups of uronic acids), anion retention (occurrence of NH_2 groups) and biological activity (energy source for microorganisms) are the other soil properties influenced by polysaccharides.

4.2.2. The Major Groups of Carbohydrates

The carbohydrates can be divided into 3 subclasses:

- Monosaccharides that are aldehyde and ketone derivatives of higher polyhydric alcohols

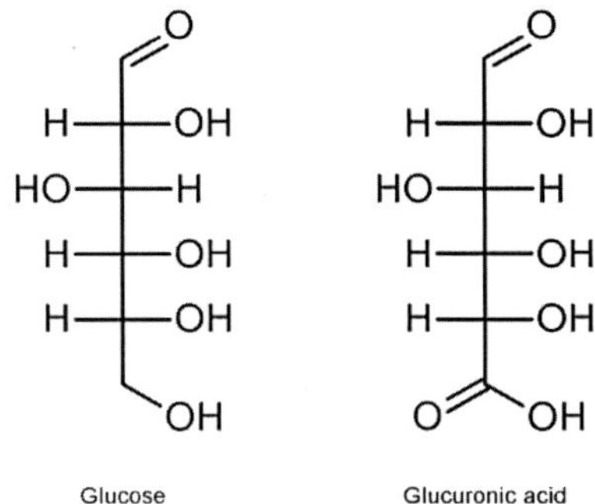

Glucose   Glucuronic acid

- Oligosaccharides that are a large group of polymeric carbohydrates comprising a relatively few monosaccharide units

- Polysaccharides that contain many monomeric units (8 or more)

The carbohydrates in soil occur as:

- Free sugars in the soil solution
- Complex polysaccharides
- Polymeric molecules of various sizes and shapes that are so strongly attached to clay and/or humus colloids

Cellulose

Cellulose is the chief component of plant cell wall. In structure, cellulose is a carbohydrate composed of glucose units bound together in a long, linear chain, by β-1,4-linkage (at carbon atoms 1 and 4 of sugar molecule). There are about 2,000 glucose units in a molecule, but it may vary from 10,000 to 15,000, depending upon the nature of the plant species. Molecular weight ranges from 200,000 to 2.4 million. Nearly 50 to 70% of carbohydrate is added in the form of cellulose.

Cellulose is broken by extra-cellular enzymes of microorganisms which cleave the β-1,4- linkage. The steps involved in the hydrolysis of cellulose to glucose have yet to be understood:

$$\text{Cellulose} \xrightarrow{\text{Cellulose hydrolysis}} \text{Cellobiose} \xrightarrow{\text{Cellobiase}} \text{Glucose}$$

Cellulose is rapidly decomposed by anaerobic and aerobic mesophilic bacteria, thermophilic bacteria, filamentous fungi, basidiomycetes, actinomycetes, protozoa, amoeba, etc.

Hemicellulose

Two types of hemicelluloses are recognized: (a) Polyuronoids (comprising repeated units of sugars and uronic acids) and (b) cellulosans (polymers of hexose or pentose sugars). While many celluloses are of these types, others have been found to contain arabinose, galactose, mannose and galacturonic acid.

Enzymatic hydrolysis of hemicelluloses results in the liberation of simple sugars or uronic acids, all of which are actively metabolized by soil flora. Frequently, the activities of these microbes give rise to new hemicelluloses of microbial origin which are recycled as part of soil organic fraction.

Starch

Starches are the reserve food of crop plants and are present mostly in seeds, fruits, tubers, roots and stems. Starches are mixtures of two structurally different polymers of glucose:

1. Amylose: Linear thread-like molecules of glucose joined by α-1,4-glucosidic linkages.
2. Amylopectins: Same α-1, 4-glucosidic linkage; but molecules are generally branched with side-chains attached through α-1, 4-glucosidic bonds.

Starches are decomposed by many microbes through the extra-cellular enzymes, amylases.

$$\text{Starch} \xrightarrow{\beta\text{-amylase}} \text{Maltose} \xrightarrow{\alpha\text{-glucosidase}} \text{Glucose}$$

This reaction stops at 6-1 branch points of amylopectin, leaving residual unhydrolyzed dextrin.

$$\text{Starch} \xrightarrow{\alpha\text{-amylase}} \text{Starch molecules of 6 glucose units}$$

Aerobic bacteria

$C_6H_{12}O_6 + 1½O_2 \rightarrow C_6H_8O_7$ (Citric acid) $+ 2H_2O$

$C_6H_{12}O_6 + 4½O_2 \rightarrow 3C_2H_2O_4$ (Oxalic acid) $+ 3H_2O$

$C_6H_{12}O_6 + 6O_2 \rightarrow 6CO_2\uparrow + 6H_2O$

Aerobic fungi

$C_2H_5OH + 2(O) \rightarrow CH_3COOH$ (Acetic acid) $+ H_2O$

$2CH_3COOH + (O) \rightarrow C_4H_6O_4$ (Succinic acid) $+ H_2O$

Aerobic yeast and bacteria

$C_6H_{12}O_6 \rightarrow 2C_3H_6O_3$ (Lactic acid)

$C_6H_{12}O_6 \rightarrow 2C_2H_5OH$ (Ethyl alcohol) $+ 2CO_2\uparrow$

$C_6H_{12}O_6 + 2(O) \rightarrow C_4H_8O_2$ (Butyric acid) $+ 2CO_2\uparrow + 2H_2O$

Starch decomposition

$(C_6H_{10}O_5)n + nH_2O \longrightarrow nC_6H_{12}O_6$ (Glucose)

Pectins

Pectins are most abundant in the intercellular layers and in the primary and secondary walls of all higher plants. They are polymers of polygalacturonic acid units in which some of the carboxyl groups are esterified with methyl groups. Pectins are broken down by two distinct enzymes: pectinesterase, which hydrolyses the methyl ester linkage to yield pectic acid and methanol; polygalacturonidase, which destroys the linkages between the galacturonic acid units of either pectin or pectic acid to release galacturonic acid.

Chitin

Chitin differs from the majority of polysaccharides in soil in that its basic unit consists of an aminosugar. Chitin is found in cell walls of lower plants and fungi, but the major source in soil is in the form of insect remains. The structure is similar to cellulose, but in place of glucose, N-acetylglucosamine serves as the repeating unit (β-1,4-glucosidic linkage). The extracellular enzyme chitinase decomposes chitin into N-acetylglucosamine.

Lignin

Lignin is the third most abundant constituent of plant tissues, found in middle lamella in close association with cell wall carbohydrates. Lignins do not occur in the plant in free state but rather as chemically combined lignin-carbohydrate complexes. They do not represent a uniform family of compounds, as individual lignins differ chemically depending on their source.

The lignin molecule contains C, H and O, but the structure is aromatic as opposed to cellulose and hemicelluloses. Lignin is a polymer built up from substituted derivatives of phenylpropane. Hydrolysis of many lignins yields structures such as vanillin, springaldehyde, coniferaldehyde and *p*-hydroxybenzaldehyde. Although lignins are extremely resistant, their gradual disappearance from soil indicates that they are capable of being metabolized. Many investigations have shown that higher fungi degrade lignin.

The contribution of sugars to soil organic matter is furnished in Table 4.1 below:

Table 4.1: Contribution of sugars to soil organic matter

Sugar	% of SOM
Hexose sugars	4 - 12
Pentose sugars	< 5
Cellulose	up to 15
Amino-sugars	2 - 6
Uronic acids	1 - 5
Others	Traces

4.3. Soil Lipids

The class of organic compounds designated as lipids represents a convenient analytical group rather than a specific type of compound. Lipids represent a diverse group of compounds ranging from relatively simple ones such as fatty acids to more complex substances like the sterols, terpenes, polynuclear hydrocarbons, chlorophyll, fats, waxes, and resins. The bulk of the soil lipids occur as the so-called fats, waxes, resins. In normal aerobic soils, the lipids are present mostly as remnants of plant and microbial tissues. About 2 to 6% of soil humus occurs in the form of fats, waxes, resins. Lipids are physiologically active. Whereas some compounds exert a depressing effect on plant growth, the others act as growth hormones. Waxes and related compounds may be responsible for the water-repellent condition of certain sands.

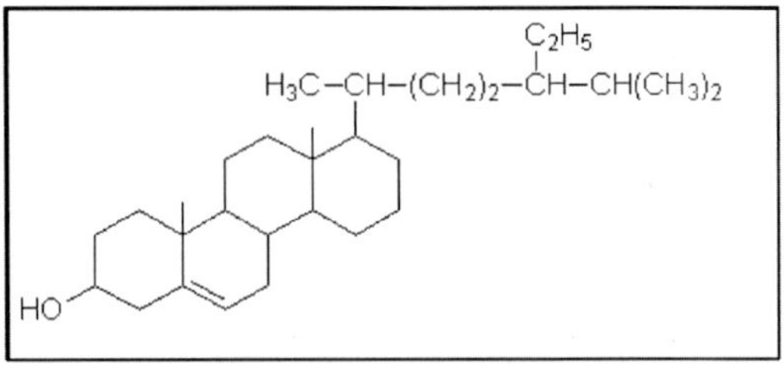

C_2H_5
$H_3C-CH-(CH_2)_2-CH-CH(CH_3)_2$
HO

5

Role of Organic Matter in Pedogenic Processes

Soil is the product of various soil forming factors acting on the parent material over a period of time. The parent material is usually devoid of organic matter and consists of variable quantities of sand, silt, clay and carbonates. High temperatures and adequate moisture promote mineral transformations, but these effects are magnified by living organisms and the organic substances they produce. The biosphere and the organic substances they produce are directly or indirectly responsible for the following:

1. Decay of organic matter by microbes leads to the formation of CO_2 which is quantitatively the most important "aggressive" weathering agent because of its tendency to form carbonic acid with water [CO_2 + $H_2O \rightarrow H_2CO_3$].
2. Many water-soluble organic substances produced in the biosphere are capable of complexing and mobilizing metal ions. Organic matter also functions as a reducing agent.
3. The uptake of certain elements by plants and microbes induces a strong driving force for reactions leading to the release of these elements into soil solution.
4. Many microbes have the ability to catalyze oxidation and reduction processes that would otherwise proceed at extremely slow pace.

The movement of organic constituents into the lower soil horizons commences at an early stage of soil development, usually in association with clay or metal ions. Worm and root channels and ped surfaces become coated with dark-coloured mixtures of humus and clay. In some soils, streaks or tongues are formed from the downward seepage of humus. Illuvial humus also appears as a coating on sand and silt particles. Localized accumulations of sesquioxides and humus are common.

In addition to leaching, organic matter can be transferred downward in soil through the action of soil animals, such as earthworms, which can completely mix soil to depths of 2 ft or more, transferring organic matter downward in this process. Burrowing animals move soil material from the deeper horizons to the surface and *vice versa.*

Organic matter may be involved in the process known as 'gley formation', a condition which occurs under impeded drainage or a high-water table and where the strong reducing conditions lead to a soil layer having light grey colour tinged with blue or green, presumably caused by Fe^{2+}.

5.1. Weathering of Rocks and Minerals

Naturally occurring organic matter plays a major role as an agent for the mobilization and transport of metal ions in soil. A series of events are involved including disintegration of parent rock material, eluviation to lower horizons and transport to plant roots. In the presence of soluble complexing agents, the total concentration of polyvalent cations in the solution phase of the soil may be several orders of magnitude higher than in their absence.

Both physical disintegration and chemical decomposition play a role in the conversion of parent rock materials into soil. Important chemical changes include hydration, oxidation, carbonization, solubilization and redeposition of breakdown products.

Biological weathering comprises the physical and chemical transformations that are brought about by living organisms and their decomposition products. For biological weathering, two types of compounds are on focus: (1) Carbonic acid which is produced from the CO_2 released during decomposition of organic matter and (2) organic chelates of microbial and higher plant origin.

Most polyvalent metal ions associated with rocks and minerals readily form soluble complexes with natural organic chelates; thus, one effect of the decomposition of mineral matter by natural chelating agents is the formation of soluble forms of various trace elements, as well as P, K, Ca and Mg.

5.2. Weathering by Simple Organic Chelates

Numerous organic compounds produced by microbes can act as solubilizers of mineral matter in nature, but organic acids appear to be the compounds mainly involved. High proportions of microbes produce organic acids. Fungi most active in dissolving silicates are those that produce the strong chelating agents citric acid or oxalic acid. A high proportion of bacteria are capable of synthesizing 2-ketogluconic acid.

5.3. Weathering by Humic and Fulvic Acids

The possibility of humic acids and related substances also participating in the chemical weathering of silicates was considered in the early history of Soil Science and evidence both for and against their involvement can be cited. Humic acids exhibited an activity of the same order as that of several simple organic compounds. On this basis, humic substances may affect metal ion mobility in environments where appreciable quantities of these substances are present in soluble forms. Because of their low molecular weights, fulvic acids may be particularly effective in dissolving silicate minerals.

5.4. Importance in Horizon Differentiation

In humid and semi-humid climates, the initial phase of soil formation leads to removal of exchangeable cations and soluble salts. The residue remaining behind consists of mixtures of silicate minerals and variable amounts of silica and sesquioxides. Subsequent weathering mediated by the decay products of organic matter results in the formation of "climax" soils, the two extremes being the laterite or Oxisol and the podzol or Spodosol. The overall reactions for these contrasting soil types are illustrated in Fig. 5.1.

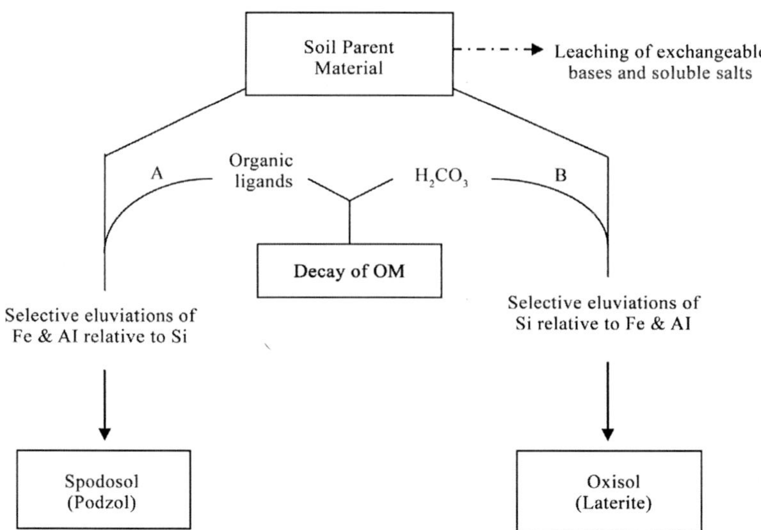

Fig. 5.1: Possible role of soil organic matter in the formation of 'climax' soils

In the case of Oxisol, the most important weathering agent is H_2CO_3, whereas in Spodosol weathering results primarily from the action of organic chelates. The latter results in the differential movement of metal ions according to their ability to form coordination complexes with organic ligands; Fe, Al and other strongly chelated elements being eluted to a greater extent than Si and similar weakly chelated ones. The downward movement of metal ions as soluble chelate complexes has been referred to as 'cheluviation' and typically occurs in Spodosols.

Spodosols have developed under climatic and biologic conditions that have resulted in the mobilization and transport of considerable quantities of Fe and Al into the subsoil. An organic-rich mineral layer of the soil, A_1, consisting largely of decomposition products of forest litter, is underlain by a light coloured eluvial horizon, A_2, which has lost substantially more Fe and Al than Si. The A_2 horizon, in turn, is underlain by a dark coloured illuvial horizon, B, in which the major accumulation products are Fe and Al and organic matter (Fig. 5.2).

The first step before the translocation of Fe, Al and organic matter can take place in a Spodosol is the replacing of a considerable portion of exchangeable cations in the upper part of the solum by H^+ ions and subsequent leaching of those exchangeable cations to the lower horizon or out of the profile. This conditioning process takes place rather rapidly on some parent materials, such as coarse textured sediments low in exchangeable cations.

5.5. Significance of Polyphenols

A popular theory presently is that polyphenols derived from the leaf litter are the main agents of podzolization. Aqueous extracts of tree litter from the forest floor of the Spodozol are capable of dissolving ferric oxide. This capability is primarily because of the polyphenols in the extracts. According to Bloomfield's concept, the solution of sesquioxides in the uppermost soil horizons results from the action of water-soluble, low molecular weight organic compounds (primarily polyphenols) leached from the overlying surface litter. Following dissolution, Fe^{3+} is reduced to Fe^{2+} by the polyphenols. It is in this form that Fe is transported down the profile.

Thus, two mechanisms are possible for the transport of Fe as Fe^{2+}: (1) Complexation followed by reduction and (2) reduction brought about under anaerobic conditions during saturation of the upper soil layers with water, followed by chelation.

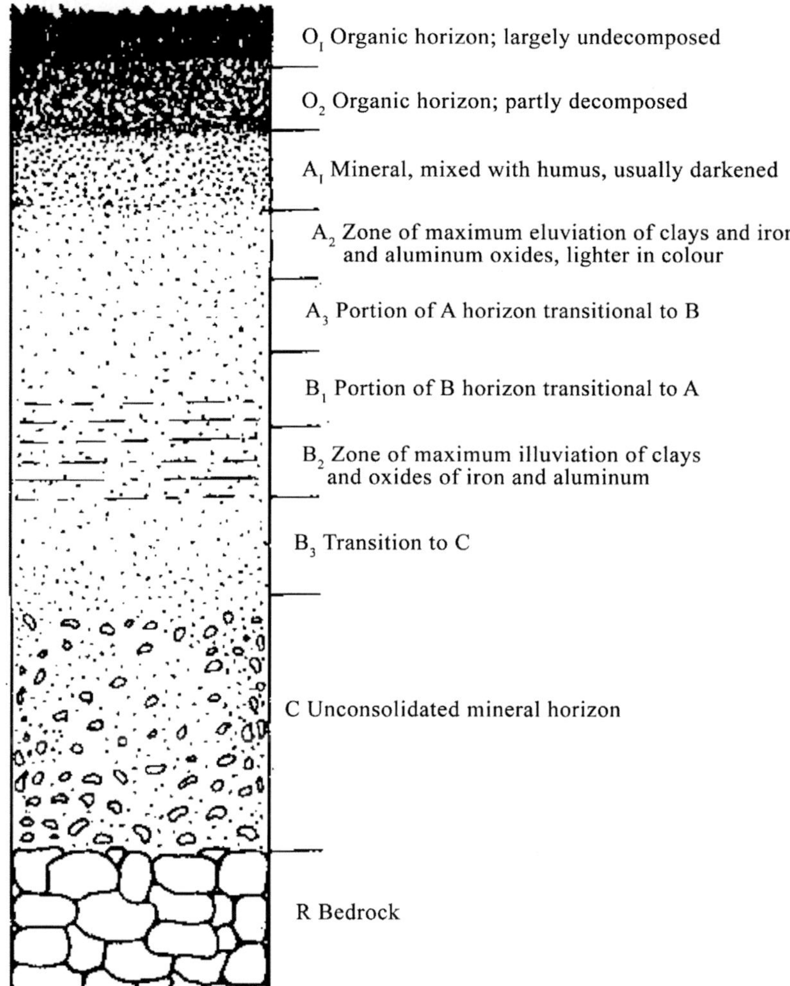

Fig. 5.2: A generalized landscape - Spodosol profile

5.6. Sesquioxide-Humus Sols

A popular theory is that the downward movement of sesquioxides in the Spodosol is caused by the formation of negatively charged sols with humic substances. Humus can carry with it from 3 to 10 times its weight of iron oxide and the different fractions of humus vary in their ability to maintain oxides in solution. A fulvic acid solution percolating down the soil profile forms complexes with Fe and Al until a critical metal-fulvic acid ratio is attained, succeeding which precipitation occurs.

5.7. Organic Matter and Soil Classification

The importance of organic matter in the comprehensive system of soil classification (7th Approximation) is credited not only to the amounts present, but also to its effect on other soil properties, especially colour. Organic matter is particularly important at the higher category levels of soil taxonomy (order, suborder, great group, subgroup).

5.8. Organic Matter and Diagnostic Surface Horizons

Organic matter content determines whether any given soil horizon can be considered "organic". If clay is absent, 20% or more organic matter must be present. When the mineral portion contains over 50% clay, at least 30% organic matter is needed. Intermediate contents of clay require proportional amounts of organic matter.

The O_1 horizon is defined as the layer where original forms of most vegetative matter are visible to the eye. The O_2 organic horizon is one in which extensive decomposition has occurred and few plant or animal parts can be recognized.

The epipedon where organic matter content is of greatest significance is in the identification of histic epipedon or peat layer. The horizon must have a thickness of 45 cm (30 cm if drained) and it must contain at least 20% organic matter if devoid of clay and 30% or more if the mineral matter contains 50% clay. Intermediate clay contents require proportional amounts of organic matter. Other restrictions are that the plough layer must contain more than 15% organic matter if clay is absent or 25% if the clay content of the mineral portion exceeds 50%. Intermediate clay contents require proportional amounts of organic matter.

The concept of mollic epipedon also centres on the properties that the organic matter confers to the top layer of the solum. With regard to organic matter, the requirements that the mollic epipedons must meet are: (1) C/N ratio must be 17 or less for virgin soils and 13 or less for cultivated soils. (2) The epipedon must contain at least 1% organic matter throughout. Other descriptions can be had from elsewhere.

5.9. Organic Matter and Diagnostic Subsurface Horizons

Organic matter is often used as a criterion for the identification of subsurface horizons, a typical example being the spodic horizon. A spodic horizon is readily recognized in the field by its dark colour and structure. The presence of organic complexes and a high CEC (attributable to the illuvial organic matter) is a further indication of a spodic horizon.

5.10. Paleohumus

Remnants of plant and animal life have been observed in buried soils (paleosols) of all ages. The organic matter, or paleohumus, is of interest in geology and pedology because of its importance as a stratigraphic marker and a key to the geological past. Thus, the occurrence of dark coloured humus zones has served as a basis for establishing the identification of buried soils from which it has been possible to obtain conclusions pertaining to climate, vegetative patterns and the morphology of former land surfaces.

6

Decomposition of Organic Residues in Soil

Soil organic matter is among the most complex materials existing in nature. In addition to the organic constituents present in undecayed plant and animal tissues, soil organic matter contains living and dead microbial cells, microbially synthesized compounds and an endless array of derivatives of these materials produced as the result of microbial activity. Soil organic matter probably contains most of the naturally occurring organic compounds. Some components of soil organic matter are no doubt distinctive to the soil environment, particularly those involving inorganic-organic complexes.

6.1. Composition of Soil Organic Matter

The composition of soil organic fraction is as follows:

1. Components of organic residues undergoing decomposition
2. Metabolic products of microorganisms utilizing organic residues as a source of energy
3. Products of secondary synthesis in the form of bacterial plasma
4. Humic substances

The first three categories listed above consist of different non-specific substances (such as proteins and its decomposition products, carbohydrates, organic acids, fats, resins, waxes, etc.) and constitute 10 – 15% of soil organic matter. Humic substances constitute 85 – 90%.

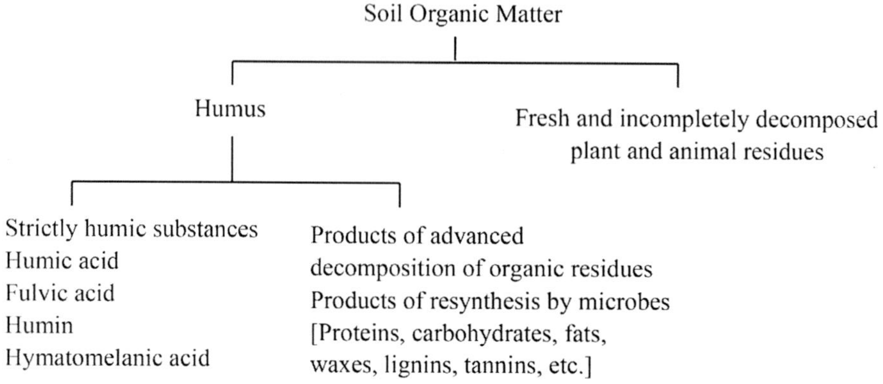

6.2. Proximate Constituents of Soil Organic Matter

Included in the extremely diverse group of compounds that constitute soil organic matter are:

- Carbohydrates and related compounds [mono- and disaccharides, celluloses, hemicelluloses, pectins, pentosans, mannans, polyuronides, uronic acids, organic acids, alcohols, hydrocarbons, aromatics]
- Proteins and their derivatives [aminoacids, amides, amino sugars, nucleoproteins, purine and pyrimidine bases]
- Lignins and their derivatives
- Fats and related substances
- Tannins and their derivatives
- Resins and terpenes

6.3. Composition of Plant Residues

The composition of plant residues that constitute soil organic matter is given in Table 6.1.

Table 6.1: The composition of proximate constituents of plant residues

S. No.	Constituent	Composition (%)
1.	Carbohydrates	
	i. Sugars, Starches	1 - 5
	ii. Celluloses	20 - 50
	iii. Hemicelluloses	10 - 20
2.	Fats, waxes and tannins	1 - 8
3.	Proteins (simple & crude)	1 - 15

6.4. Decomposition of Soil Organic Matter

The organic fraction of the soil is a complex of substances whose composition is determined in part by the plant and animal residues added to the soil, and to a greater extent, by the transformations of these substrates through biological, physical and chemical means. As a result, a vast array of compounds is present: those representing the components of organic residues undergoing decomposition and the metabolic by-products of the microorganisms utilizing such compounds as a source of energy. Thus, in soil we find two major decomposition processes taking place simultaneously:

1. Breakdown of organic substrates, both fresh and native
2. Synthesis of humic substances

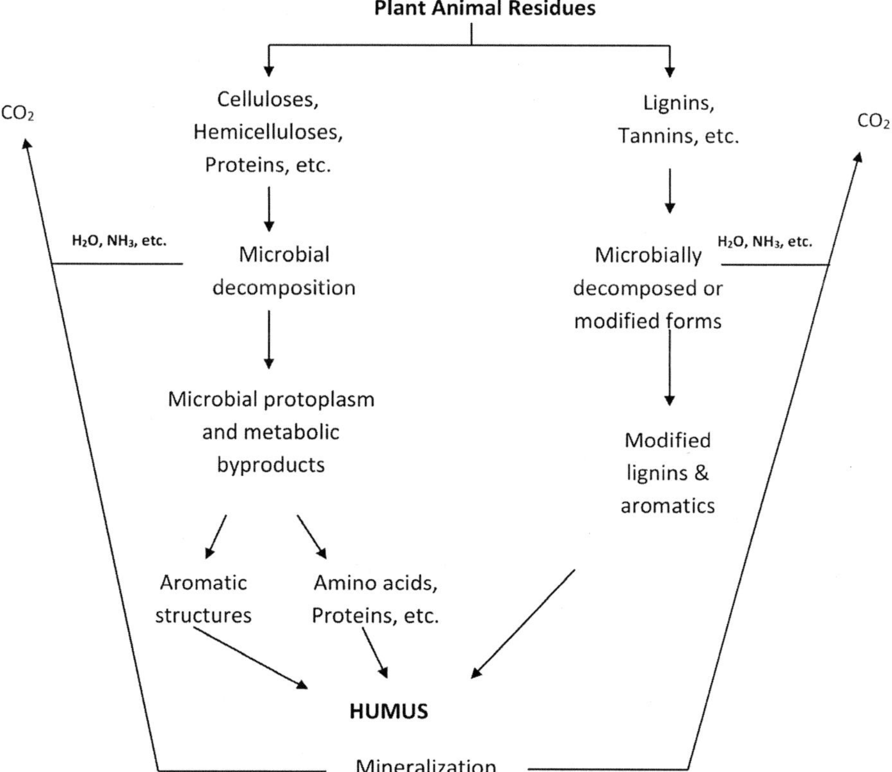

Fig. 6.1: Organic matter decomposition and formation of humic substances in soil

The outline of general changes is as follows:

1. Compounds of plant tissues
   a. Those compounds that decompose with difficulty [Fats, oils, lignins, resins, waxes, etc.]
   b. Those that decompose easily [Celluloses, starch, sugars, proteins, etc.]
2. Complex, intermediate compounds of decomposition
   a. Those that are resistant to microbial action [Resins, waxes, fats, lignins, oils, etc.]
   b. Those that decompose readily [Aminoacids, amides, alcohols, carbohydrates, aldehydes, etc.]
3. Decomposition products
   a. Resistant compounds [Humus]
   b. Simple end products [CO_2, H_2O, NO_3, SO_4, PO_4, Ca compounds, etc.

Most soils contain a substantial pool of N in organic forms that are so large that they are not amenable for rapid direct uptake. Only after having been cleaved into smaller units by extracellular enzymes, these larger forms of organic N become available by and large. Enzymatic cleavage of organic N is quite often compared with total depolymerisation of protein into amino acids. Still, it is likely that small peptides are also products of depolymerisation and there could be a suite of other N-containing compounds that are produced through enzymatic cleavage.

Enzymatic cleavage of endogenous organic matter results in production of small peptides (generally < 500 Da) and amino acids in approximately equal amounts. Hexosamines, muramic acid and diaminopimelic acid account for less than 1% of the products of enzymatic cleavage, while at least 2% is accounted for by nucleobases + nucleosides and headgroups of membrane lipids.

6.5. Climate Change and Decomposition of Soil Organic Matter

Plant roots are the primary source of soil organic carbon and critically support the growth and activities of microbes in the rhizosphere. Factors of climate change may, however, alter root-microbial interactions and affect carbon dynamics in the rhizosphere.

Studies have shown that while warming and reduction in precipitation suppressed microbial decomposition of root litter irrespective of the presence of roots, increase in precipitation invigorated litter decomposition only in the absence of roots, indicating that plant competition for water restraints the activities of saprophytic microbes.

Though presence of roots increased microbial biomass, it reduced microbial activities such as respiration, carbon cycling enzymes and litter decomposition, indicating that roots exert differential effects on microbes through altering carbon or water availability. In addition, nitrogen input significantly reduced microbial biomass and microbial activities (respiration). Thus, alterations in soil moisture induced by climate change drivers critically modulate root controls over microbial decomposition in soil.

In a nutshell, warming-enhanced plant water utilization, combined with nitrogen-induced suppression of microbes, may provide a unique mechanism through which moderate increases in precipitation, warming and nitrogen input interactively suppress microbial decomposition, thereby facilitating short-term soil carbon sequestration in the arid and semi- arid regions.

7

Nutrient Transformations in Soil Nitrogen, Phosphorus and Sulphur

Soil is the fundamental resource in the agricultural production systems and monitoring its fertility is critical to the sustainability of agriculture. Plant nutrients undergo different transformation processes through which they are converted from unavailable to available forms. Nutrient cycling thus involves the transformation and availability of nutrients and is mediated by microbial action. Understanding the dynamics of nutrient and soil organic matter transformations requires an understanding of the biological, chemical, and physical processes involved.

7.1. Soil Nitrogen

The quantity of nitrogen in surface soils generally ranges from 0.02 to 0.25 per cent and is closely related to the amount of soil organic matter of which nitrogen makes up approximately 5 per cent. Approximately 98 per cent of nitrogen in soils occurs in organic form. Inorganic nitrogen rarely exceeds 2 to 3 per cent of the total soil nitrogen.

7.1.1. The Nitrogen Cycle

The availability of nitrogen is of prime importance to growing plants since they are dependent on an adequate supply of nitrate and ammonia for synthesis of their nitrogenous constituents. Animals, in turn, depend on plants for their energy. When plant and animal remains are added to soils, their nitrogenous constituents undergo numerous transformations, many of which are biologically opposed. The net result is that only a small proportion of the total N in soil is present in an available form at any point of time. The cycle of nitrogen as it exists in nature is illustrated in Fig. 7.1.

The behaviour of nitrogen in soils resides in the numerous and varied reactions that occur between organic and inorganic forms of nitrogen. The nitrogen cycle that operates in soil is distinct from the overall cycle of nitrogen in nature, but interfaces with it.

A key feature of soil nitrogen cycle is the biological turnover of nitrogen between mineral and organic forms through the opposing processes of *mineralization* and *immobilization*. The latter leads to incorporation of nitrogen into microbial tissues, the 'active fraction of organic matter'. While much of this newly immobilized nitrogen is recycled through mineralization, some is converted into stable humus forms. Fig. 7.2 furnishes the overall process.

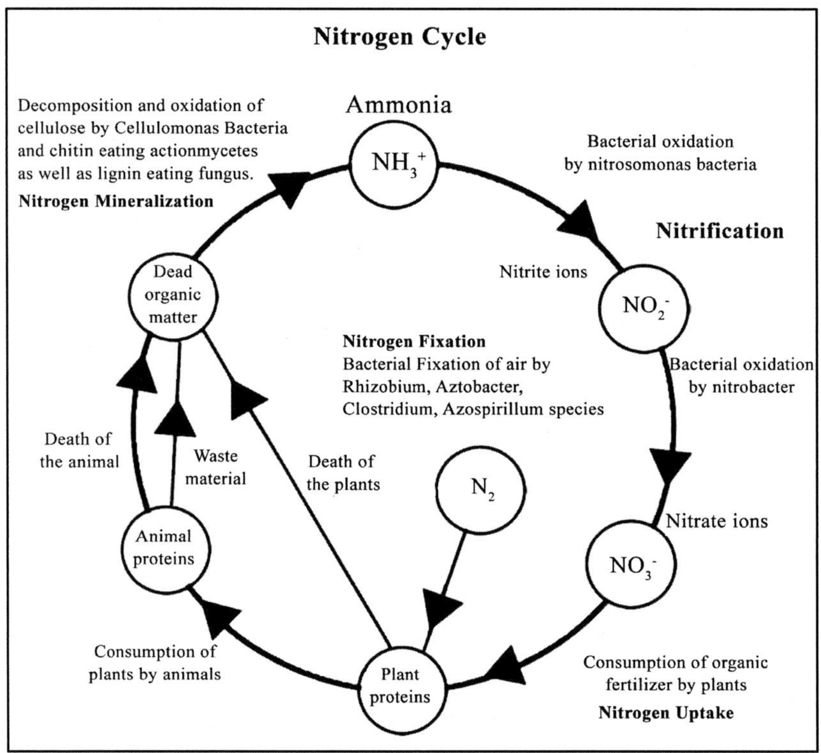

Fig. 7.1: The soil nitrogen cycle in nature

Thus, the decay of plant and animal residues by microbes results in the formation of mineral forms of nitrogen (NH_4^+ and NO_3^-) and assimilation of part of the carbon (and mineralized nitrogen) into microbial tissue (Reaction A). A part of the native humus undergoes a similar fate (Reaction B). Subsequent turn over through mineralization-immobilization results in incorporation of nitrogen into stable humus forms (Reaction C). Stabilization of nitrogen may also occur through the reaction of partial decay products of lignin with nitrogenous constituents (Reaction D).

Except under unusual circumstances, both mineralization *(the process of conversion of organic form of an element into simple inorganic forms by microbial activity)* and immobilization *(the process by which simple inorganic nitrogen compounds are changed to organic combinations in microbial tissues or plant tissues)* always function in soil, but in opposite directions. Thus, measurements for changes in mineral nitrogen levels (NH_4^+ and NO_3^-) merely provide an indication of differences in magnitude of the two opposing processes. A decrease in mineral nitrogen levels with time indicates *net immobilization*; an increase indicates *net mineralization.*

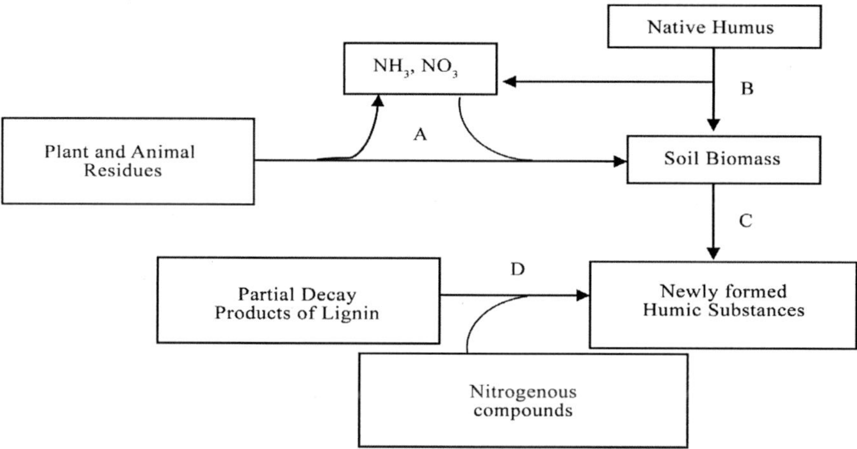

Fig. 7.2: Reactions depicting incorporation of organic forms of N into stable humus

7.1.2. Mineralization

The mineralization of nitrogenous compounds takes place in two distinct steps:

1. Ammonification: Organic N $\rightarrow$ NH_4^+
2. Nitrification: $NH_4^+ \rightarrow NO_3^-$

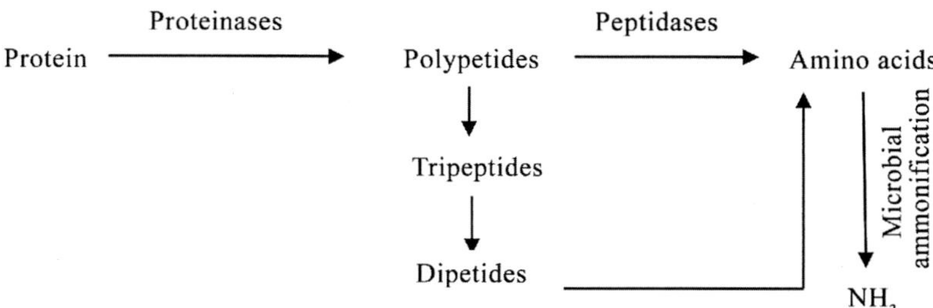

7.1.3. Ammonification

Ammonification takes place through:

1. ***Oxidative Deamination***

   $HOOC.CH_2.CH_2.CH.NH_2.COOH$ (Glutamic acid) + $O_2 \rightarrow COOH.CH_2.CH_2.COOH$ (Succinic acid) + $NH_3 + CO_2$

2. ***Hydrolytic Deamination***

   $R.CH.NH_2.COOH$ (Protein) + $H.OH \rightarrow R.CH.OH.COOH$ (Organic acid) + NH_3

   $R.CH.NH_2.COOH + H.OH \rightarrow R.CH_2OH + CO_2 + NH_3$

   $R.CH.NH_2.COOH + H.OH \rightarrow R.CHO + HCOOH + NH_3$

3. ***Decarboxylation***

   $CH_3.CH.NH_2.COOH$ (Alanine) + $H_2 \rightarrow H_3C.CH_3$ (Ethane) + $NH_3 + CO_2$

4. ***Desaturative Deamination***

   $R.CH_2.CH.NH_2.COOH \rightarrow R.CH = CH.COOH + NH_3$

7.1.4. Nitrification

Nitrification is the process by which ammonia is converted to nitrate. It is brought about by two specialized groups of aerobic autotrophic bacteria which derive energy from oxidation of specific inorganic compounds.

Nitrosomonas $NH_4 + 1\frac{1}{2}O_2 \rightarrow NO_2^- + 2H^+ + H_2O$

Nitrobacter $NO_2^- + \frac{1}{2}O_2 \rightarrow NO_3^-$

7.1.5. Denitrification

Denitrification is the biochemical reduction of nitrate or nitrite to gaseous nitrogen or oxides of nitrogen. Two steps are involved in the process by which microbes utilize NO_3: (1) Nitrate assimilation or assimilatory nitrate reduction which is the biological reduction of NO_3 to ammonia and (2) Nitrate respiration or dissimilatory nitrate reduction in which NO_3 is used as a terminal electron acceptor.

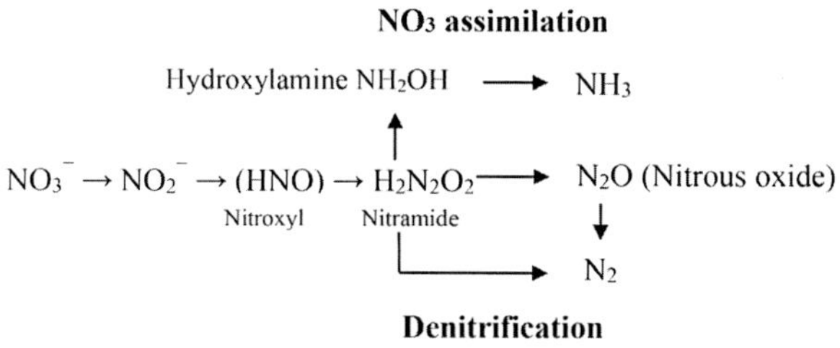

NO_3 assimilation

Hydroxylamine NH_2OH → NH_3

↑

$NO_3^- \rightarrow NO_2^- \rightarrow (HNO) \rightarrow H_2N_2O_2 \rightarrow$ N_2O (Nitrous oxide)

Nitroxyl Nitramide ↓

→ N_2

Denitrification

In the denitrification process, two distinct groups of anaerobes are involved. The first group includes those which reduce NO_3 (nitrates) to NO_2 (nitrites) and allows the product to accumulate. The second group completely reduces NO_2 to N_2 and nitric or nitrous acids. Important denitrifiers (autotrophs) are *Micrococcus denitrificans* and *Thiobacillus denitrificans*.

Fungi play a key role in the nitrogen cycle and diverse fungi (mostly species from the order Hypocreales, particularly *Fusarium oxysporum* and *Trichoderma* spp.) are known to reduce nitrate or nitrite to gaseous nitrogen oxides (such as nitric oxide and nitrous oxide) and dinitrogen *via* denitrification and to ammonium *via* ammonia fermentation (fungal dissimilatory nitrate reduction to ammonium). These processes could significantly lead to the emission of N_2O from soils and remove nitrogen from nitrate and nitrite-contaminated environments.

7.2. Soil Phosphorus

The major sources of phosphorus in soil are: (1) the organic compounds present in plant and animal residues and products of microbial synthesis (such as phytin, phospholipids, nucleoproteins, nucleic acids, phosphorylated sugars and related compounds) and (2) the inorganic compounds in which phosphorus is combined largely with calcium, magnesium, iron, aluminium and clay minerals.

Soil organic matter accounts for 25 to 80% of total phosphorus in soils. The major proportion of phosphorus in soil humus is present as phytin or its derivatives (40 to 80%), nucleic acids (0 to 10%), together with phospholipids (0.3%).

Transformations of phosphorus in which microbes are involved include: (1) the decomposition and subsequent mineralization of phosphorus bound in organic residues of plant and animal origin; (2) the altering of the solubilities of inorganic compounds of phosphorus; (3) the immobilization or the assimilation of phosphorus into microbial tissues with ultimate deposition in the soil humus; and (4) the possible participation in the oxidation and reduction of inorganic phosphorus compounds. The various phosphorus components, inputs and losses in soil are presented in Fig. 7.3. The phosphorus transformations in soil are furnished in Fig. 7.4

7.2.1. The Phosphorus Cycle

The **phosphorus cycle** is a biogeochemical cycle which describes the path of phosphorus through the lithosphere, hydrosphere and biosphere. In stark contrast to many other biogeochemical cycles, the atmosphere does not play a significant role in the movement of phosphorus, because phosphorus-based compounds are usually solids at the normal ranges of temperature and pressure on Earth.

Phosphorus occurs in nature amply as part of the orthophosphate ion (PO_4^{3-}, consisting of a phosphorus atom and 4 oxygen atoms. On land, phosphorus occurs mostly in rocks and minerals. Phosphorus-rich deposits have generally formed in the ocean or from guano, and with time, geologic processes brought ocean sediments to land. Upon weathering, rocks and minerals release phosphorus in a soluble form to be taken up and transformed into organic compounds by plants. When herbivores consume plants, phosphorus is incorporated into their tissues or gets excreted. After death and decay of animals and plants, phosphorus is returned to the soil where a large portion of it is converted into insoluble compounds. Runoff may carry along a small portion of the phosphorus back to the ocean. Generally, with time (thousands of years), soils become deficient in phosphorus leading to the retrogression of ecosystem.

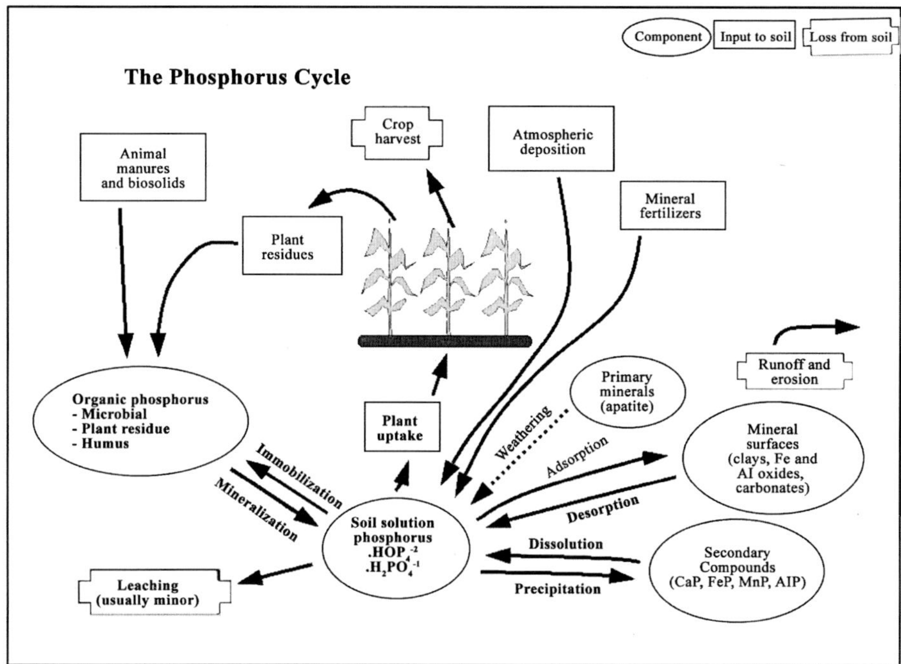

Fig. 7.3: The phosphorus cycle

Though phosphates move quickly through plants and animals, the processes of their movement through the soil or ocean are very slow, making the overall phosphorus cycle one of the slowest biogeochemical cycles. The global phosphorus cycle encompasses four major processes:

- tectonic uplift and exposure of phosphorus-bearing rocks such as apatite to surface weathering;
- physical erosion and chemical and biological weathering of phosphorus-bearing rocks to provide dissolved and particulate phosphorus to soils, lakes and rivers;
- riverine and subsurface transportation of phosphorus to various lakes and run-off to the ocean;
- sedimentation of particulate phosphorus (e.g., phosphorus associated with organic matter and oxide/carbonate minerals) and eventually burial in marine sediments (this process can also occur in lakes and rivers).

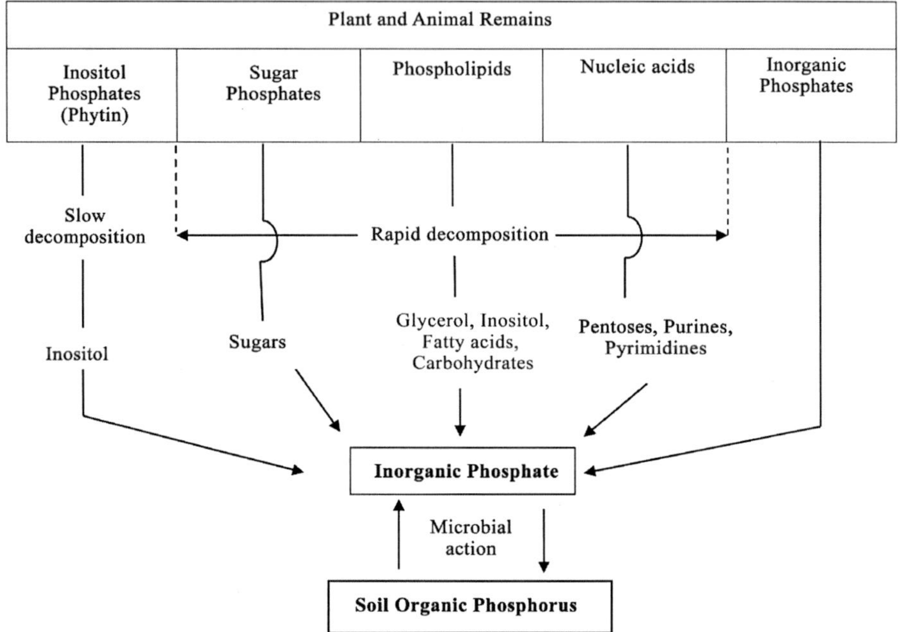

Fig. 7.4: Phosphorus transformations in soil

In terrestrial systems, bioavailable phosphorus chiefly comes from weathering of phosphorus-containing rocks. Apatite, the most abundant primary phosphorus-mineral in the crust, can be dissolved by acids produced by soil microbes and fungi, or by other chemical weathering reactions. This dissolved phosphorus which is bioavailable to plants and other terrestrial organisms is returned to the soil after their decay.

Phosphorus retention by soil minerals (e.g., adsorption onto iron and aluminium oxyhydroxides in acidic soils and precipitation onto calcite in neutral-to-calcareous soils) is usually the most important processes in controlling terrestrial bioavailability of phosphorus in the mineral soil. This process can result in low concentrations of dissolved phosphorus in soil solution.

Soil phosphorus is usually transported either to the lakes to be buried in lake sediments or to the ocean *via* river runoff. In surface seawater, dissolved inorganic phosphorus, mainly orthophosphate (PO_4^{3-}), is assimilated by phytoplankton and gets converted into organic phosphorus compounds. Cell lysis of phytoplankton releases cellular dissolved inorganic and organic phosphorus into the surrounding environment.

Enzymes synthesized by bacteria and phytoplankton hydrolyze some of the organic phosphorus compounds which are subsequently assimilated. A vast majority of phosphorus is remineralized within the water column and roughly 1% of associated phosphorus carried to the deep sea by the falling particles is removed from the ocean reservoir and buried in sediments. A series of processes augment sediment pore water phosphorus concentrations. These processes include:

(i) microbial respiration of organic matter in sediments,

(ii) microbial reduction and dissolution of iron and manganese oxides with subsequent release of associated phosphorus, which connects the phosphorus cycle to the iron cycle, and

(iii) abiotic reduction of iron oxides by hydrogen sulphide and liberation of iron- associated phosphorus.

7.2.2. Phosphorus Transformations

Phosphorus is second only to nitrogen as an inorganic nutrient required by plants and microorganisms. This element has a unique metabolic role in all living systems as it is intimately involved in the energy reactions of ATP. Phosphate derivatives also are important structural components of nucleic acids, coenzymes and phospholipids, as well as being involved in almost all the significant metabolic pathways pertinent to life.

1. Forms of Phosphorus in Soil

Soil phosphorus occurs in different organic and inorganic forms. In the surface horizon of most soils, organic forms of phosphorus account for 20 to 90% of the total phosphorus. This percentage is lower than that for nitrogen that is almost all in organic forms in soil.

Soil organic phosphorus occurs as monoester [$(RO)PO_3H_2$] and diesters [$(RO)(R'O)PO_2H$] and their chemical diversity is in the organic moieties (R and R'). However, less than 50% of the organic phosphorus in most soils can be accounted for in known compounds. The principal forms are inositol phosphates, nucleic acids and phospholipids.

The source of phosphorus in non-cultivated soils is in the parent material from which the soil is formed. The biological reactions that lead to stable organic matter accumulation require available phosphorus and when low concentrations of phosphorus in the parent material occur, these reactions will be restricted. The accumulation of organic matter in many soils is considered to be inhibited primarily because of a lack of available phosphorus.

To avoid loss of phosphorus from natural ecosystems, a tight cycling of phosphorus due to biological activity occurs. For example, in the Brazilian rainforest where virtually all of the phosphorus exists in living and dead organic matter and the underlying mineral soil contains extremely low levels of available phosphorus, the phosphorus cycle is virtually closed. Most of the plant phosphorus is recycled as a result of microbial breakdown of litter and organic debris.

Phosphorus is often applied to the soil as a fertilizer in agricultural production systems. Most of the phosphorus applied, often as high as 90%, is not taken up by the crop but is retained in insoluble or fixed mineral forms. Further additions of fertilizer phosphorus are required in subsequent years to maintain high crop yields. Phosphorus applications repeatedly in excess of crop removal inevitably result in its accumulation. Many cultivated soils in industrial countries have significantly greater phosphorus concentrations than their uncultivated counterparts. In areas where there are high concentrations of animal production units, the manure disposal/use options are often restricted to land application and excessive levels of soil phosphorus that have environmental implications can develop.

2. Mineralization and Immobilization of Organic Phosphorus

Turnover of phosphorus *via* mineralization and immobilization reactions are similar to that of nitrogen in that both processes take place simultaneously.

Organic phosphorus ⟷ Inorganic phosphate

When crop residues are returned to the soil, net immobilization will occur if the organic carbon/organic phosphorus ratio is 300 or more. Net mineralization will occur if the ratio is 200 or less. If the organic carbon/organic phosphorus ratio is between 200 and 300, neither a gain or loss of inorganic phosphate occurs.

Because turnover of organic phosphorus is a biological process, factors that affect biological activity such as soil moisture, pH, and energy supply will also affect the mineralization and immobilization reactions.

There are several specific enzymes associated with the mineralization of organic phosphorus compounds in soil. These belong to three main classes: the phosphomonoesterases, the phosphodiesterases and inorganic pyrophosphatase. The action of phosphomonoesterases involves the removal of an organic moiety from phosphoric acid as follows:

$$R - O - \underset{\displaystyle OH}{\overset{\displaystyle O}{\overset{\|}{\underset{|}{P}}}} - OH + H_2O \rightarrow ROH + HO - \underset{\displaystyle OH}{\overset{\displaystyle O}{\overset{\|}{\underset{|}{P}}}} - OH$$

The substrate *p*-nitrophenyl phosphate is commonly used to determine phosphomonoesterase activity in soil. The R group in *p*-nitrophenol is a chromophore which can be quantitatively extracted from soil and detected at very low concentrations. Inositol phosphates and nucleotides are examples of molecules that are stripped of their phosphate group by phosphomonoesterases.

The phosphomonoesterases frequently exhibit two pH optima in soil. In acid soils, optimum activity is observed in the slightly acid range and the enzyme responsible is called acid phosphatase. In alkaline soils, the major portion of activity occurs at approximately pH 11 and the enzyme is called *alkaline phosphatase.*

Phosphodiesterases catalyze the reaction in which a phosphodiester is hydrolyzed to form a phosphomonoester and an alcohol. Phosphodiesterases degrade phospholipids and nucleic acids.

$$R_1 - O - \underset{\displaystyle OH}{\overset{\displaystyle O}{\overset{\|}{\underset{|}{P}}}} - OR_2 + H_2O \rightarrow R_1OH + HO - \underset{\displaystyle OH}{\overset{\displaystyle O}{\overset{\|}{\underset{|}{P}}}} - OR_2$$

Inorganic pyrophosphatase is an enzyme that is ordinarily associated with energy-related reactions and is ubiquitous in nature. However, superphosphate fertilizer contains 50% or more of its total phosphorus as pyrophosphate. For plants to take up phosphorus from the pyrophosphate molecule it must be split.

$$HO - \underset{\displaystyle OH}{\overset{\displaystyle O}{\overset{\|}{\underset{|}{P}}}} - O - \underset{\displaystyle OH}{\overset{\displaystyle O}{\overset{\|}{\underset{|}{P}}}} - OH + H_2O \rightarrow 2HO - \underset{\displaystyle OH}{\overset{\displaystyle O}{\overset{\|}{\underset{|}{P}}}} - OH$$

Pyrophosphate Phosphoric acid [H_3PO_4]

3. Oxidation-Reduction Reactions

Phosphorus can exist in a number of oxidation states ranging from -3 in phosphine, PH_3, to + 5 in orthophosphate. Phosphite (PO_2^{3-}) and hypophosphite (PO_2^{3-}) can both be oxidized in soil *via* biological reactions, as determined by observing that the reactions cease if a biological inhibitor, as toluene, is added. However, soils do not normally contain the phosphite or hypophosphite molecules and hence the significance of the oxidation reactions is rather mute.

The reverse process - reduction of phosphate to phosphite and hypophosphite - has also been shown to take place when soils are incubated anaerobically in a mannitol-ammonium dihydrogen phosphate ($NH_4H_2PO_4$) medium. When nitrate or sulphate is added to the soil, the reduction is inhibited as the nitrate and sulphate are readily used as terminal electron acceptors. Unlike nitrogen or sulphur, there is still controversy concerning whether the ultimate reduced phosphorus compound may be a gas, in this case, phosphine. Some scientists claim that this reduction step is thermodynamically impossible while others have published data indicating phosphine evolution from soil. Although this reaction is also probably of limited practical significance, a definite answer as to whether it occurs or not in soil will continue to be debated.

7.3. Soil Sulphur

Sulphur is present in soils in both inorganic and organic forms. In well-leached surface soils, much of the sulphur is combined with organic matter. Soils of humid temperate regions frequently have 50 to 500 ppm of water-soluble sulphates and 100 to 1500 ppm of total sulphate. Sulphur is added to soils in rain and snowfall, through many commercial fertilizers and irrigation waters.

7.3.1. The Sulphur Cycle

The **biogeochemical sulphur cycle** consists of various processes that together enable the movement of sulphur through different reservoirs like the atmosphere, biosphere, and lithosphere. Sulphur is an important mineral in living organisms which is present in biomolecules like proteins. It is the tenth most abundant element on the Earth's surface and is found in different forms in the rocks and soil.

The sulphur cycle consists of both the terrestrial and atmospheric processes and the element in different oxidation states moves through various ecosystems on Earth, affecting both the biological and geological process. Because of its chemical structure, sulphur remains stable in multiple oxidation states which enables the formation of a diverse group of organic and inorganic compounds. All the reserves of sulphur are bound to the lithosphere, from which they are steadily released by weathering processes. Sulphur compounds formed during the cycle can act as either oxidants or as reductants, depending on their oxidation state.

Microorganisms are vital in the sulphur cycle as they have specialized enzyme systems and mechanisms to form different sulphur compounds. Sulphur found in nature might even exist in a combined form with other elements like phosphorus and nitrogen. The concentration of sulphur on Earth is influenced by human activities like the addition of sulphur-containing fertilizers to the soil. Unlike inorganic phosphorus compounds, sulphur molecules are usually more soluble in water which increases their availability to plants and microorganisms in soil. The following are the steps involved in the biogeochemical sulphur cycle:

1. In the atmosphere

- Sulphur in the atmosphere is mostly present in the form of SO_2. Most of the SO_2 in the atmosphere arises from anthropogenic activities like combustion of fossil fuel.
- Natural processes like volcanic eruptions also play an essential role in increasing the SO_2 concentration in the atmosphere.
- SO_2 is followed by H_2S, which represents another important gas that is present in the atmosphere.
- The primary source of H_2S in the atmosphere is the gas released by microbial action on dead decaying living organisms.
- The microbes present both in the terrestrial and the aquatic habitats decompose organic and inorganic forms anaerobically.
- The anaerobic decomposition results in the release of H_2S, which then oxidizes in the air to form SO_2.
- The sulphate-reducing microorganisms involved in the decomposition process are usually anaerobic where they produce H_2S from the oxidized form of sulphur.
- Some of the sulphate-reducing bacteria include *Desulfovibrio desulfuricans, Desulfovibrio vulgaris, Thermodesulfovibrio yellowstonii, Desulfotomaculum nigrificans* and *Desulfobacula toluolica*.

2. In the Biosphere

- The sulphur enters the biosphere in one of two ways: from the atmosphere or weathering of rocks. In either of the two ways, the sulphur eventually makes it to the soil and then to the ocean.
- The sulphur in the atmosphere aids in the cloud formation by increasing the number of cloud droplets and the droplet size to decrease.

- The sulphur particles, also called *aerosols*, fallout from the atmosphere to the biosphere.
- The SO_2 present in the atmosphere reaches the biosphere as the gas dissolves in the rainwater to form weak droplets of sulphuric acid.
- Besides, chemical weathering in the pedogenesis process also results in the movement of sulphur from rocks to the soil and the water.
- Weathering also releases sulphur into the air as some of it is converted into sulphate.

3. Plant and Animal Uptake

- Once it reaches the terrestrial and aquatic biosphere, sulphur is consumed by plants and microorganisms.
- A group of bacteria called the *green sulphur bacteria* act as photoautotrophic bacteria and utilize sulphur as a form of energy.
- Other microorganisms in the soil also aid in making sulphur available to plants so that it can be taken up with water from the soil.
- The sulphur taken up by living beings is then used up to form biomolecules like proteins and nucleotides.
- In the ocean ecosystem, chemoautotrophic microorganisms utilize sulphur to produce organic compounds in the form of sulphates.

4. Lithification and Release

- The sulphur in the biosphere then circulates through the food chain as the consumers feed on producers and then reach the microbial chains.
- The sulphur that does not circulate falls into the depths of terrestrial and marine habitats and remains there in the combined form (FeS) in rocks.
- The sulphur in the food chain then undergoes the decomposition process, converting the sulphate into sulphides so that they can be returned back to the atmosphere.
- Sulphur reducing bacteria act on the organic forms of sulphur to form inorganic forms like hydrogen sulphide (H_2S), which is further reduced to S.
- The sulphur in the lithosphere also releases back to the atmosphere as a result of volcanic activity.

Plants build up their S-containing constituents primarily through the assimilation of sulphates from soil and small amounts through absorption of S-aminoacids or direct assimilation of SO_2 form the atmosphere. Animals, in turn, utilize plants or their residues, converting the S compounds into organic molecules. When plant and animal remains are returned to the soil, they are decomposed by microbes with the liberation of S in inorganic forms which may be oxidised to sulphates or reduced to H_2S by many diverse organisms.

The sulphur cycle in soil can be divided into several distinct phases: (1) mineralization – the decomposition of large organic molecules into simple inorganic S compounds; (2) immobilization – the assimilation of these compounds into microbial tissues; (3) oxidation - the conversion of inorganic forms of S (sulphides, thiosulphates, polythionates and elemental S) to sulphate and (4) reductions – the conversion of sulphates and other oxidized forms of S to H_2S (Fig. 7.5).

7.3.2. Sulphur Transformations

Sulphur is an essential macronutrient for members of the plant and animal kingdom. For many years, the study of sulphur was neglected as compared to the other major nutrients because most crops did not respond to sulphur fertilization. However, with greater use of high analysis fertilizers, crop responses to sulphur are becoming more common. Also, the concern over the environmental issues of acid rain, as well as sulphur dioxide pollution, has stimulated a new awareness on the importance of sulphur in our environment.

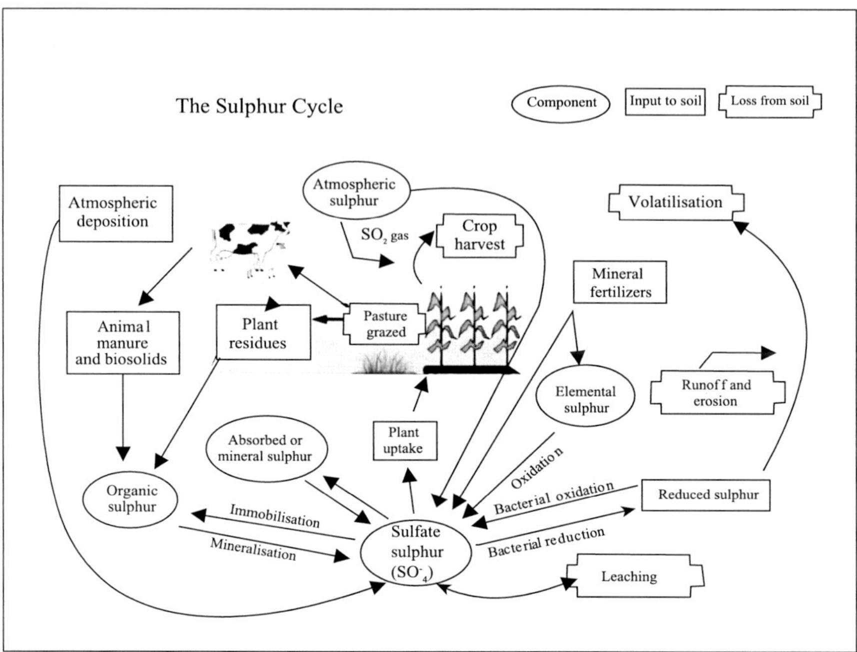

Fig. 7.5: The sulphur cycle

Ninety-five percent or more of the total sulphur in soils is in organic forms. Exceptions to this occur in limited areas where saline, acid sulphate, and other soils dominated by sulphur-containing minerals are located. The organic sulphur in soils is often divided into two fractions, the sulphate ester compounds ($R\text{-}O\text{-}SO_3H$) and the carbon-bonded sulphur compounds (primarily sulphur amino acids).

Sulphur transformations in soil are considered to result primarily from the activity of microorganisms and can be classified as follows:

1. Mineralization - the breakdown of organic containing sulphur compounds to form mineral sulphate.
2. Immobilization - the conversion of mineral sulphate to organic sulphur compounds.
3. Oxidation - the conversion of inorganic sulphur compounds of lower oxidation state to a higher oxidation state.
4. Reduction - the conversion of inorganic sulphur compounds of higher oxidation state to a lower oxidation state.

The transformations of sulphur resemble those of nitrogen in many ways. Both the elements exist in soil primarily in the organic state and undergo mineralization and immobilization reactions. Sulphur is oxidized from sulphide to sulphate (somewhat analogous to oxidation of ammonium to nitrate) and the requirements of the soil for the reaction to proceed are similar. Also, both nitrogen and sulphur can be reduced to gaseous forms and lost from the soil.

7.3.3. Enzyme Reactions in Soil Involving Sulphur Compounds

Since much of the soil organic sulphur is present as sulphate esters, aryl and alkyl sulfatase enzymes play a key role in sulphur mineralization. The overall reaction can be written as

$$R - OSO_3^- + H_2O \longrightarrow R - OH + H^+ + SO_4^{2-}$$

Arylsulfatases in soil are assayed by measuring the amount of *p*-nitrophenol released from *p*- nitrophenyl sulphate added to soil and incubated for 1 to 2 hours. The activity of arylsulfatase varies according to soil type, soil depth, organic matter content, season and climate.

The two major sulphur-containing amino acids, cysteine and methionine, also undergo enzyme-catalyzed transformations in soil. First, however, oxidation of cysteine to cystine (the disulphide form of the amino acid) rapidly occurs in soil as this reaction can be catalyzed by trace amounts of a number of metal ions. An enzyme called *cystathionine lyase* acts upon cystine to form a disulphide called *thiocysteine*. Thiocysteine then reacts with a free sulfhydryl group to form hydrogen sulphide (H_2S). The series of reactions are:

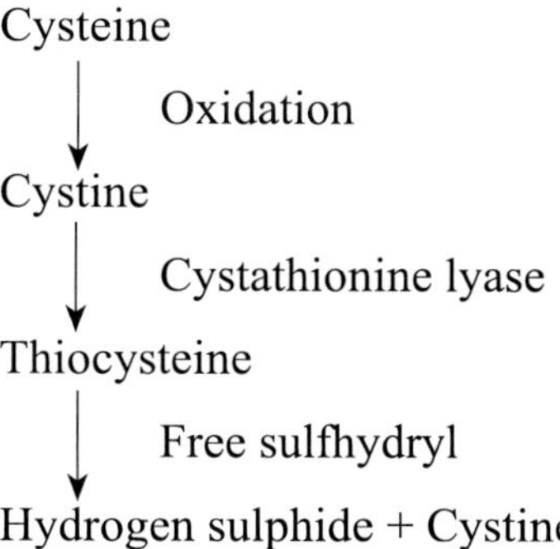

Inspection of the chemical structures of cysteine and thiocysteine illustrates how the hydrogen sulphide may be formed during the sequence of reactions described.

$HSSCH_2CH(NH_2)COOH + HSCH_2CH(NH_2)COOH \rightarrow H_2S\uparrow + (SCH_2CH(NH_2)COOH)_2$

Thiocysteine Cysteine Cystine

In environments that are neither highly aerobic nor anaerobic, both cysteine and cystine (which contains a free sulfhydryl) may be present. Field experiments have shown that losses of hydrogen sulphide are more likely to occur during the beginning or a period of waterlogging than later after a more strongly anaerobic condition has developed, i.e., when sulphate reduction would be expected to be the dominant mechanism of hydrogen sulphide production. The series of reactions just described may explain this reaction.

One other enzyme engaged in the sulphur cycle has been detected and characterized in soil. This enzyme *rhodanese* belongs to the transferase class of enzymes. It catalyzes the formation of thiocyanide from thiosulphate and cyanide according to the following reaction.

$$S_2O_3^{2-} + CN^- \longrightarrow SCN^- + SO_3^{2-}$$

Rhodanese activity is found in a large variety of soils. Both thiosulphate and tetrathionate are intermediary products during the oxidation of elemental sulphur to sulphate and the rhodanese-catalyzed reaction may be involved in the further metabolism of these compounds in soil.

8

Biochemistry of Humus Formation

The synthesis of humic substances is the least understood and the most intriguing aspect of humus chemistry. This has been one of the intensely-researched subjects for a long time. Formation of humic substances takes several pathways during the decay of plant and animal remains in soil and the four important ones are shown in Fig. 8.1.

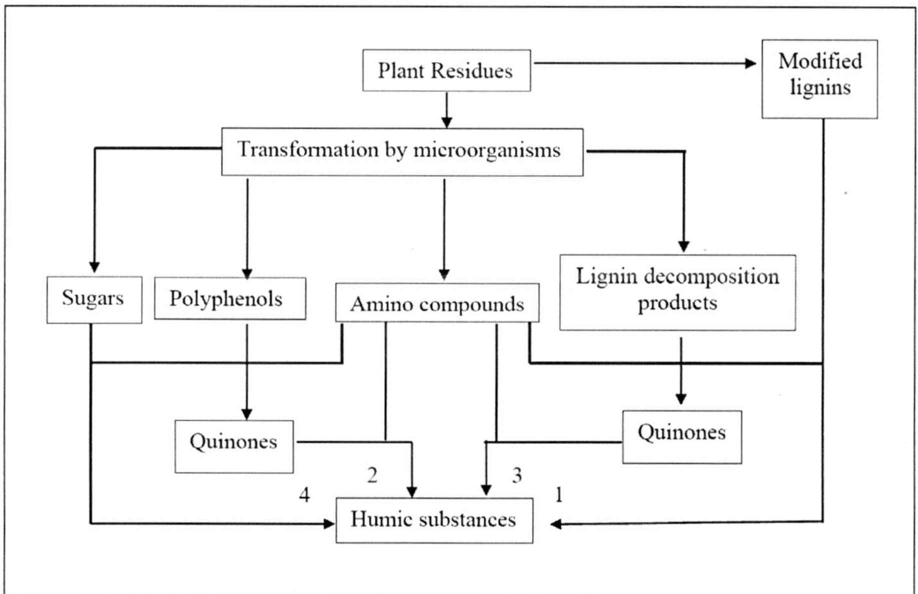

Fig. 8.1: Mechanisms for the formation of soil humic substances

The classical theory by Waksman advocated humic substances as modified lignins (Pathway 1), but a vast majority of current crop of investigators favour a mechanism involving quinones (Pathways 2 and 3). Including sugar-amine condensation pathway (Pathway 4), all the four pathways must be considered as likely mechanisms for the synthesis of humic and fulvic acids in nature. It is absolutely possible that these four pathways may operate in all soils, though not to the same extent or in the same order of importance. Lignin pathway may be dominant in poorly drained soils and wet sediments (swamps, etc.), whereas synthesis from polyphenols may be of great significance in certain forest soils. In terrestrial surface soils of frequent and sharp fluctuations in temperature, moisture and irradiation under a harsh continental climate, humus synthesis by sugar-amine condensation may be the favoured pathway.

8.1. The Lignin Theory

Since many years it was thought that humic substances were derivatives of lignin (Pathway 1). As per this theory, microorganisms utilize lignin incompletely and the residuum becomes an integral part of the humus. Modifications in lignin include loss of methoxyl (OCH_3) groups with the generation of o-hydroxyphenols and oxidation of aliphatic side chains to form COOH groups. The modified material obtained thus undergoes further unknown changes to yield first humic acids and then fulvic acids. This pathway, illustrated in Fig. 8.2, is exemplified by Waksman's lignin-protein theory. Waksman cited the following evidence in support of the lignin theory of humic acid formation:

- Both lignin and humic acid are decomposed by a vast majority of fungi and bacteria with considerable difficulty.
- Both lignin and humic acid are partly soluble in alcohol and pyridine.
- Both lignin and humic acid are soluble in alkali and are precipitated by acids.
- Both lignin and humic acid contain OCH_3 groups.
- Both lignin and humic acid are acidic in nature.
- When lignins are warmed with aqueous alkali, they are transformed into methoxyl- containing humic acids.
- Humic acids have properties similar to oxidized lignins.

Although microorganisms attack lignin less easily than other plant components, mechanisms do exist in nature for its complete aerobic decomposition. If not, undecomposed plant residues would accumulate on soil surface and the soil organic matter content would gradually increase until CO_2 was depleted from the atmosphere. In normal, aerobic soils lignin may be broken into low-molecular-weight products before humus synthesis. Further, the fungi which degrade lignin are not normally found in excessively wet sediments. It is, therefore, logical to presume that modified lignins may contribute considerably to the humus of peat, lake sediments, and poorly drained soils.

8.2. The Polyphenol Theory

In pathway 3 also lignin plays a significant role in humus synthesis, but in a different way. Here, phenolic aldehydes and acids released from lignin during microbial attack undergo enzymatic conversion to quinones, which polymerize in the presence or absence of amino compounds to form humus-like macromolecules. Pathway 2 is in a way similar to pathway 3, except that the polyphenols are synthesized by microorganisms from non-lignin carbon sources [e.g., cellulose]. The polyphenols are then oxidized enzymatically to quinones and converted into humic substances.

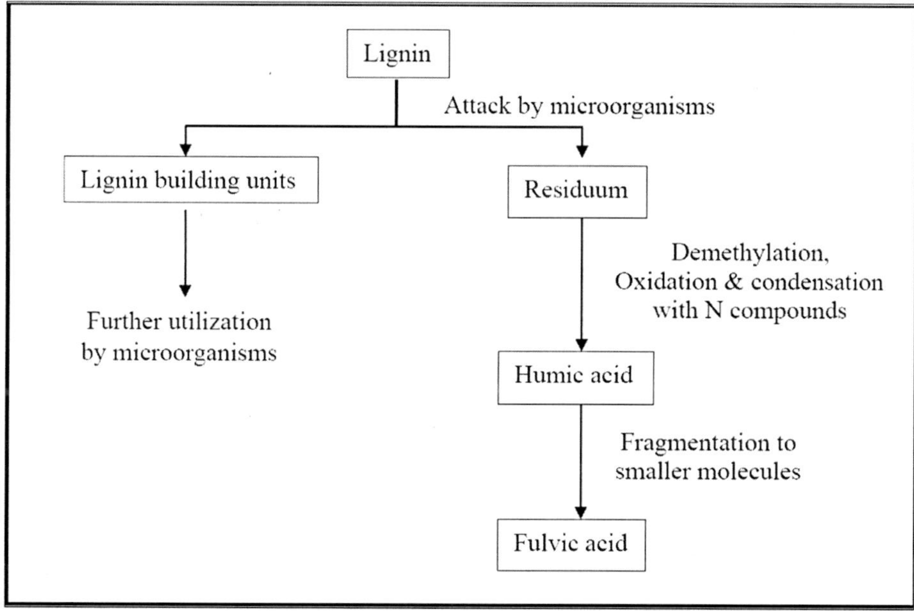

Fig. 8.2: Schematic representation of lignin theory of humus

The classical lignin-protein theory of Waksman is now considered obsolete by contemporary investigators. By current concepts, quinones of lignin origin and quinones synthesized by microorganisms are the prime building blocks from which humic substances are formed. The formation of brown-coloured substances by reactions involving quinones is a well-known phenomenon and not an uncommon event that takes place in melanine formation, such as in the flesh of ripe fruits and vegetables following mechanical injury and during seed coat formation. The possible sources of phenols for humus synthesis include lignin, microorganisms, uncombined phenols in plants, glycosides and tannins.

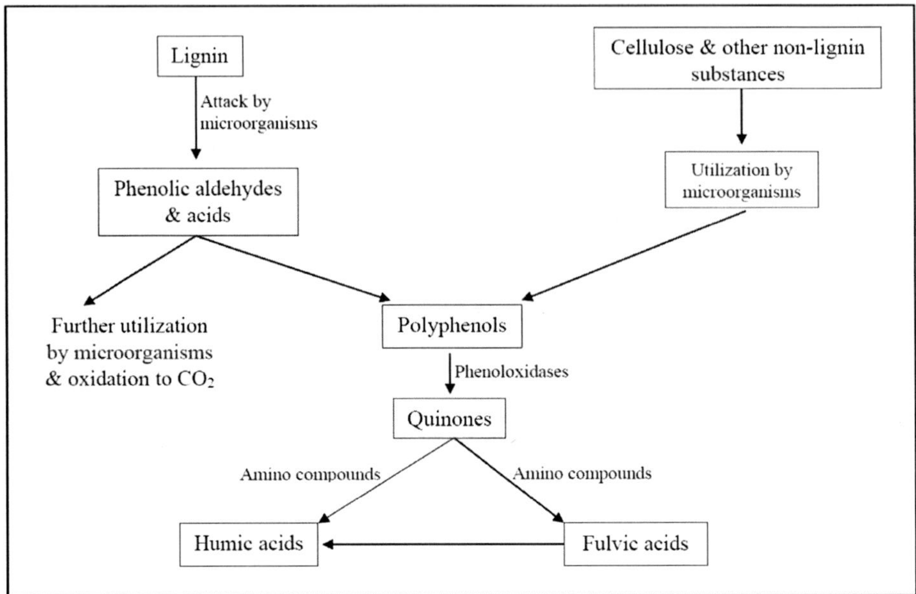

Fig. 8.3: Schematic representation of polyphenol theory of humus formation

Flaig's concept of humus formation is:

- Freed of its linkage with cellulose during decomposition of plant residues, lignin is subjected to oxidative splitting with the formation of primary structural units (derivatives of phenyl propane).
- The side-chains of the lignin-building units are oxidized, demethylation occurs and the resulting polyphenols are converted to quinones by polyphenol oxidase enzymes.
- Quinones formed from the lignin (and from other sources) react with N-containing compounds to form dark-coloured polymers.

Kononova emphasized the role of microorganisms as a source of polyphenols and concluded that humic substances were being formed by cellulose-decomposing myxobacteria prior to lignin decomposition. The three stages culminating in the formation of humic substances were postulated as:

Stage 1: Fungi attack simple carbohydrates and parts of protein and cellulose in the medullary rays, cambium and cortex of plants residues.

Stage 2: Aerobic myxobacteria decompose cellulose of the xylem. Polyphenols synthesized by the myxobacteria are oxidized to quinones by the enzyme polyphenol oxidase. Quinones subsequently react with nitrogen compounds to form brown humic substances.

Stage 3: Lignin is decomposed. Phenols released during decay are also part of source materials for humus synthesis.

8.3. Sugar-amine Condensation Theory

The notion that humus is formed from sugars dates back to the early days of humus chemistry. According to this notion, reducing sugars and amino acids which are the by-products of microbial metabolism undergo non-enzymatic polymerization to form brown nitrogenous polymers. A major objection to this theory is that the reaction proceeds rather too slowly under the normal conditions of soil temperature. However, drastic changes in the soil environment (such as freezing and thawing, wetting and drying), together with the intermixing of reactants with mineral material having catalytic properties, may promote condensation. An attractive feature of the theory is that the reactants (sugars, amino acids, etc.) are produced in abundance through the activities of microorganisms.

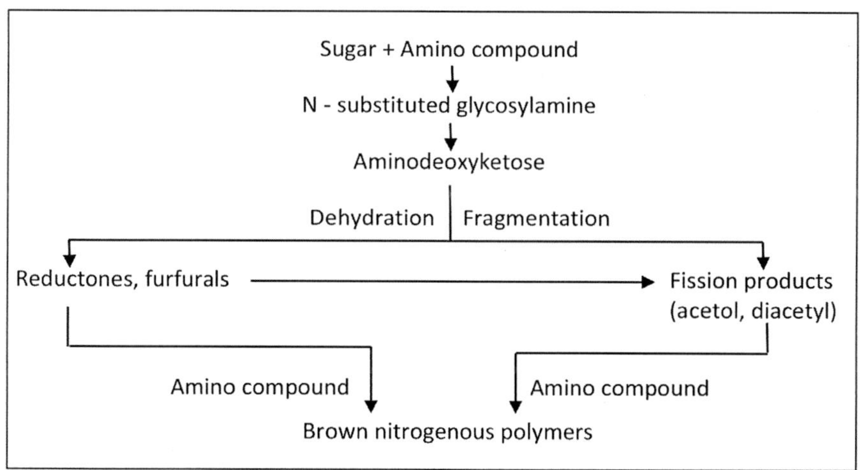

Fig. 8.4: Schematic representation of sugar-amine condensation theory of humus formation

The initial reaction in sugar-amine condensation involves addition of the amine to the aldehyde group of the sugar to form the N-substituted glycosylamine which subsequently undergoes Amadori rearrangement to form the N-substituted-1-amino-deoxy-2-ketose. This is subject to: (1) fragmentation, with the liberation of the amine and formation of 3-carbon chain aldehydes and ketones, such as acetol, glyceraldehyde and dihydroxyacetone, (2) loss of two water molecules to form reductones and (3) loss of three water molecules to form hydroxymethyl furfural. All these highly reactive compounds readily polymerize in the presence of amino compounds to form brown-coloured products of unknown composition.

8.4. Humus

Humus is a dark-coloured (brown or dark-brown) complex mixture of amorphous and colloidal substances arising from modified plant materials and synthesized microbial tissues.

Humic Substances

Humic acids: The fraction of humic substances that is not soluble in water under acidic conditions (pH < 2) but is soluble at higher pH values is humic acid. Humic acids are the major extractable components of soil humic substances and are dark brown to black in colour.

Fulvic acids: The fraction of humic substances that is soluble in water under all pH conditions is fulvic acid. Fulvic acids remain in solution after humic acids are removed by acidification. They are light yellow to yellow-brown in colour.

Humin: The fraction of humic substances that is not soluble in water at any pH value and in alkali is humin. Humin is black in colour.

8.5. Characteristics of Humic Substances

The following are the major elemental composition (%) of humic substances:

Element	Humic acid (%)	Fulvic acid (%)
C	56.2	45.7
H	04.7	05.4
N	03.2	02.1
S	00.8	01.9
O	35.5	40.8

The following are the functional groups of humic substances: carboxyl, phenolic OH, alcoholic OH, carbonyl, methoxyl, ketone groups. Such functional groups are important in ion exchange reactions.

Many present investigators firmly believe that all dark colored humic substances are part of a system of closely related, but not completely identical, high molecular weight polymers. The differences between humic acids and fulvic acids can be explained by variations in molecular weight, numbers of functional groups (carboxyl, phenolic OH) and extent of polymerization.

The postulated relationships are depicted in Fig. 8.5, in which it can be seen that carbon and oxygen contents, acidity and degree of polymerization all change systematically with increasing molecular weight.

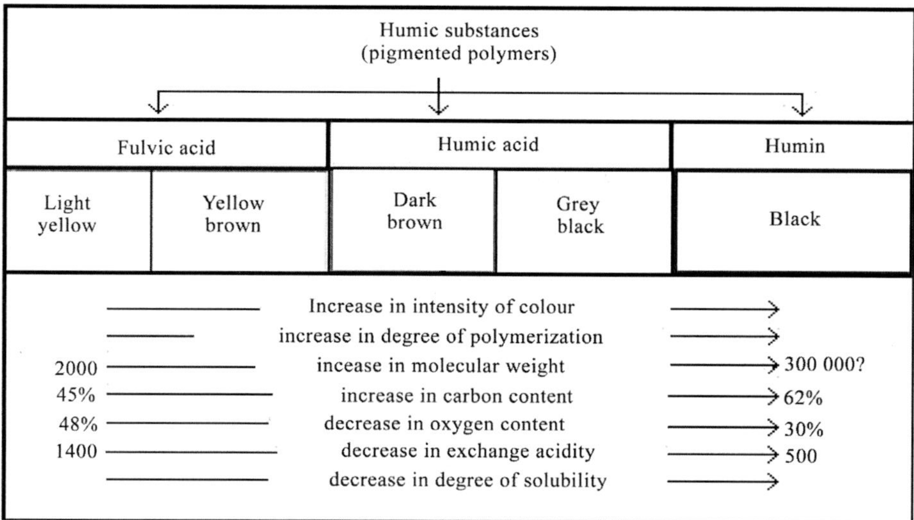

Fig. 8.5: Properties of humic substances

The low molecular weight fulvic acids have higher oxygen but lower carbon contents than the high molecular weight humic acids. Fulvic acids contain more functional groups acidic in nature, particularly COOH. The total acidities of fulvic acids (900 - 1400 meq 100 g^{-1}) are considerably higher than for humic acids (400 - 870 meq 100 g^{-1}). Another important difference is that while the oxygen in fulvic acids can be accounted for in known functional groups (COOH, OH, C=O), a high proportion of the oxygen in humic acids seems to occur as a structural component of the nucleus.

9

Trace Metal Interactions with Humic Substances

Practically every aspect of the chemistry of heavy metals in soil is related to the formation of complexes with organic matter. While monovalent cations (Na^+, K^+, etc.) are held mainly by simple cation exchange through the formation of salts with COOH groups (RCOONa, RCOOK), multivalent cations (Cu^{2+}, Zn^{2+}, Mn^{2+}, etc.) form coordinate linkages with organic molecules.

A schematic diagram showing organic matter reactions involving metal ions in soils is given in Fig. 9.1. The metals present in the solution phase as charged species and as soluble metal-organic complexes (MChe), are shown to be influenced by the activities of higher plants and microbes, both of which serve as sources of water-soluble ligands for complex formation; some metals are held in insoluble organic complexes and are non-leachable and relatively unavailable to plants.

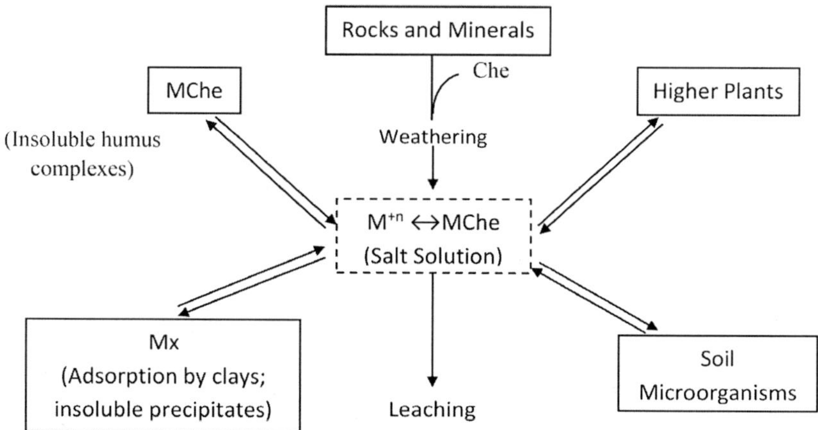

Fig. 9.1: Organic matter reactions involving metal ions

The quantity of any given polyvalent cation in soil solution (often in parts per billion range) at any point of time is normally trivial in comparison with the total amount held by clay and humus colloids, or as precipitates. However, from the stand point of plant nutrition, the soluble cations are of greatest importance. With reference to complex formation and plant nutrition, the metals can be listed in the following three groups:

1. Those which are essential to plants but are not bound in coordinate compounds.

   Included are monovalent cations such as Na^+ and K^+ and divalent cations as Ca^{2+} and Mg^{2+}.

2. Those metals which are essential to plants and which form coordinate linkages with organic ligands. They include nearly all of the metals in the first transition series, including Cu^{2+}, Zn^{2+}, Mn^{2+} and Co^{2+} as well as Mo of the second transition series.

3. Those without known functions in plants but accumulate in the environment. Included in this group are Cd^{2+}, Pb^{2+}, Hg^{2+}, chromium, gold, uranium and vanadium.

9.1. Chemistry of Chelates and Complexes

A metal ion in aqueous contains attached water molecules oriented in such a way that the negative (oxygen) end of the water dipole is directed towards the positively charged metal ion. A complex arises when water molecules surrounding the metal ion are replaced by other molecules or ions, with the formation of a coordination compound. The compound which combines with the metal ion is commonly referred to as the *ligand.*

A covalent bond consists of a pair of electrons shared by two atoms and occupying two orbitals, one of each atom. Typically, a coordinate complex arises owing to the fact that the outer electron shell of the metal ion is not completely filled and, therefore, can accept additional pairs of electrons from whatever atoms that have a pair of electrons ready for sharing. Examples of groupings contained in organic compounds which have unshared pairs of electrons and which can form coordinate linkages with metal ions, are shown by structures I to IV:

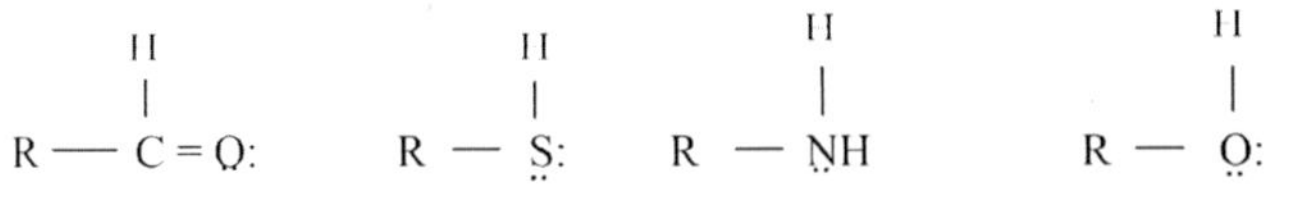

H	H	H	H
|	|	|	|
R — C = O:	R — S:	R — NH	R — O:
I	II	III	IV

In Soil Science, the two terms *complex formation* and *chelation* have been used interchangeably. However, with respect to the nature of bonding, a distinction may be drawn between complex formation and chelation. Complex formation is the reaction between a metal ion and ligands through electron-pair sharing. The resultant is a *metal coordination compound.* The metal ion is the electron-pair acceptor, while the ligand is the electron-pair donor. The metal ion as the central ion and the organic ions coordinated around it constitute the first coordination sphere. Since they are surrounded by water molecules, hydrated metal ions in solution are also considered as complexes with water.

The number of ligands bonded to a central atom in a definite geometry is called the 'coordination number'. Some of the organic ligands can bind the metal ion with more than one donor functional group. This type of bonding results in a heterocyclic ring, called a *chelate* ring.

The process of formation of a chelate ring is called *chelation* (Greek; *chele* means crab's claw). If only one ligand molecule is involved in the formation of a chelate, the compound is called as a *monodentate chelate* or complex; if two ligand molecules form a chelate with a metal, the complex compound is a *bidentate*. Depending on the number of ligands participating in the formation of chelate, we can as well have *tridentates*, *tetradentates*, and *pentadentates,* with more than one ligand imparting high stability to compound.

Almost any metal atom can serve as an acceptor atom, including K^+, Li^+, Na^+, Ag^+ and Au^+. A well-known complex compound with a monovalent ion is potassium ferrocyanide. A ligand can be an anion (Cl^- or $R = CH_2 - COO^-$) or a neutral molecule (NH_3). The complexes formed can be cations, anions, or neutral molecules.

Several naturally occurring soil organic acids are capable of complexing metal ions. The reaction occurs chiefly with the transition metals (Al, Fe, Cu, Zn, and Mn) and is often treated as a special case of an adsorption process. Such kind of adsorption is quite different from the regular Coulombic adsorption of cations in an electric double layer. The metal ion "adsorbed" in a complex reaction cannot be exchanged rapidly in the traditional way of cation exchange reaction.

The bonds in a complex compound are covalent bonds which are stronger bonds than the electrostatic bonds in cation exchange reactions. However, an exchange of the complexed metal is still possible, but such an exchange depends on many factors, e.g., soil pH, affinity of the metals for the ligand and stability of the complexes. The exchange will take place more rapidly between transition metals. For example, an exchange of complexed Al by free Fe occurs easier than by Na^+ ions. The exchange by Na^+ ions is very difficult, since the Na^+ ion cannot occupy the centre position of the Al^{3+} ion in the complex.

9.2. Soil Organic Compounds

The organic compounds in soil, capable of interacting with inorganic substances, can be differentiated into *non-humified* and *humified organic* compounds. The non-humified materials are biochemicals produced by the living organisms and may range from simple aliphatic acids to complex aromatic, heterocyclic acids and miscellaneous biochemicals. They include formic acid, acetic acid, amino acid, benzoic acid, citric acid, oxalic acid, tannic acid, tartaric acid, vanillic acid and miscellaneous compounds, such as phytic acid, chlorophyll, sugars, sugar acid and polysaccharides. These compounds have been released into the soil during the decomposition process of plant and animal residue, whereas many of them may have been released as root exudates. Some are intermediate products of plant and microbial metabolism and the others are products of oxidative degradation of organic matter.

The rhizosphere contains microorganisms that synthesize large amounts of the organic acids and other biochemical chelating agents. The availability of Fe in alkaline soils is reportedly caused by organic chelating agents formed in the rhizosphere by microorganisms or secreted by plant roots.

Most of these acids in soil generally have very low concentrations. For example, the amount of acetic acid is between 0.7 and 1.0 mmol 100 g^{-1} soil. In general, the concentration of organic acids in the soil solution is normally in the range of 1×10^{-3} to 1×10^{-4} mol L^{-1}. Higher concentrations have usually been detected in the rhizosphere and in the leachates from forest canopies. Though amino acids have been detected in higher concentrations than the other organic acids, their role as chelating agents is only secondary to the other biochemicals.

9.3. Metal - Organic Complex Reactions

Humic matter can form both soluble and insoluble complexes with metal ions. Some of the factors affecting solubility of metal organic complexes are: (1) pH, (2) presence of salts or cations, (3) dissociation of functional groups and (4) saturation of binding sites. Metal complexes of fulvic acid are, in general, more soluble than those of humic acids, perhaps because of the lower molecular weights and higher solubility of fulvic acids in water.

The maximum chelating (binding) capacity of humic matter equals its *total acidity* (= amounts of H^+ from COOH + phenolic OH groups). A total acidity of 1000 meq 100 g^{-1} corresponds to 90 mg of Al^{3+} g^{-1} of humic acid, bonded by the functional groups. The problems associated with solubility of metal-humic acid complexes are usually described in terms of stability constants. Depending on the stability constants of the complex compound, they can be soluble or insoluble in water. Assume that the following reaction occurs:

$M^{2+} + 2HA = MA_2 + 2H^+$

where M = metal ion with valence 2+, HA = humic acid, MA_2 = metal - humic acid complex

According to the Mass Action law, the equilibrium constant K of equation above is then:

$K = [(MA_2)(H^+)^2] / [(M^{2+})(HA)^2]$

Upon taking the log, this equation changes to:

$\log K = \log [(MA_2)(H^+)^2] / [(M^{2+})(HA)^2]$

If the activities of HA and MA_2 are taken as unity, then:

$K = \log (H^+)^2 / (M^{2+})$

Or,

$\log K = 2 \log (H^+) - \log (M^{2+})$

where log K is called the 'stability constant'. It determines the solubility of the metal complexes. The greater the value of log K, the greater is the stability of the complex compound. The effect of pH in increasing the stability constants of metal-fulvic acid complexes is given below:

Stability constants	Cu-FA	Zn-FA	Mg-FA
Log K (pH 3.5)	7.15	5.40	3.42
Log K (pH 5.5)	8.26	5.73	4.06

Factors affecting stability constants

- Basicity of ligand - stronger the base the more stable the complex is (correlated to ability to accept proton)
- Increasing the number of metal chelate rings/ ligand increases stability
- Size of chelate ring (5-6 numbered rings best)
- Nature of metal and ligand - the more the bonds tend toward electrostatic bonds the more stable the complex is.

9.4. Significance of Chelation Reactions in Soil

The formation of metal-organic complexes would have the following effects in soils, sediments and natural waters:

1. Metal ions that would ordinarily convert to insoluble precipitates at pH values found in productive agricultural soils would be maintained in solution. It is well known that many biochemicals synthesized by microbes, including aminoacids and simple aliphatic acids form soluble complexes with metal ions. The formation of soluble complexes with lead and other toxic heavy metals is of concern from the standpoint of their transport to lakes and streams.

2. Organic complexing agents may influence the availability of trace elements to higher plants, as well as to soil micro- and macro-faunal organisms.

3. Under certain circumstances, the concentration of a metal ion may be reduced to a non-toxic level through complexation. This would be particularly true when the metal-organic complex has low solubility, such as is the case with complexes of humic acid and other high-molecular weight components of the organic matter.

4. Chelation process has a significant role to play in the chemical weathering of rocks and minerals. Lichens (as well as bacteria and fungi) bring about disintegration of rock surfaces to which they are attached, through the production of organic chelating agents.

5. Natural complexing agents are of tremendous importance in the transportation and concentration of metals in a variety of commercially important biogenic deposits, such as peat and coal. For instance, toxic levels of Cu in peat have been attributed to a series of events that include its uptake from the surrounding mineral soil by plant roots, translocation to leaves, incorporation into the humus of the forest floor following leaf fall and migration to the swamp in seepage waters as soluble organic complexes.

6. Complexing agents of various types function as carriers of heavy metals in natural waters.

7. The interaction of Al^{3+} with organic matter may be of considerable importance in controlling soil solution levels of Al^{3+} in acid soils.

10

Reactive Functional Groups of Humic Substances

The chemistry of humic substances is the least comprehended in the field of Soil Science. The many important functions of humus will remain obscure if the structural chemistry of the humic and fulvic acids is not known. A variety of functional groups have been reported in humic substances. However, considerable disagreement still exists in respect of the exact amounts present and, in some cases, proof of existence is lacking.

Some important structural groups of organic molecules are given below:

- Amino $-NH_2$
- Amine R. NH_2
- Amide R.CO.NH_2
- Alcohol R.CH_2.OH
- Aldehyde R.CHO
- Carboxyl R.COOH
- Carboxylate R.COO-
- Enol R.CH.CH.OH
- Ketone R.CO.R'
- Keto acid R.CO.COOH
- Imine R.CH.NH
- Imino =NH
- Ether R.CH_2.O.CH_2.R'
- Ester R.COO.R'
- Anhydride R.CO.O.CO.R'

- Quinone

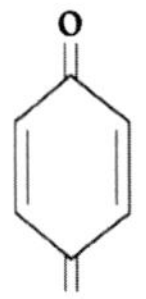

- Hydroxyquinone

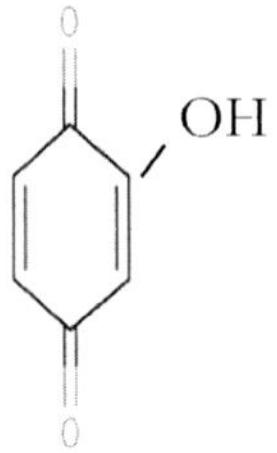

10.1. Elemental Content of Humic Substances

The major elements in humic and fulvic acids are carbon and oxygen. The carbon content of humic acids ranges from 50 to 60%; oxygen content varies from 30 to 35%. Fulvic acids have lower carbon (40 to 50%), but higher oxygen (44 to 50%) contents. Percentage of hydrogen varies from 4 to 6% and sulphur from 0 to 2% (Table 10.1).

Table 10.1: Elemental content of humic and fulvic acids

Element	Dry and ash-free basis (%)	
	Fulvic acids	Humic acids
C	40 - 50	50 - 60
O	44 – 50	30 - 35
H	4 – 6	4 – 6
N	< 1 - 3	2 - 6
S	0 - 2	0 - 2

10.2. Methods of Functional Group Analysis

Most methods used for determining reactive groups in humic materials are founded on the acidic properties of the group involved. Because of the inherent complexities in the nature of humic substances, the acidities of the various groups may overlap.

10.2.1. Total Acidity (COOH plus Phenolic- and / or Enolic-OH)

The sample is allowed to react with excess of $Ba(OH)_2$, and the unused base is titrated with standard acid:

$$2HA + Ba(OH)_2 \rightarrow BaA_2 + 2H_2O$$

The method used by Schnitzer (1972) is as follows: To between 50 and 100 mg of humic preparation in a 125-mL ground-glass stoppered Erlenmeyer flask, 20 mL of 0.2*N* $Ba(OH)_2$ solution is added. Simultaneously, a blank is set up with 20 mL of 0.2*N* $Ba(OH)_2$ only. The air in each flask is displaced by N_2, the flasks are stoppered carefully and shaken for 24 hours at room temperature. The suspension is filtered, the residue is washed thoroughly with CO_2- free distilled water and the filtrate plus washings are titrated potentiometrically with standard 0.5*N* HCl to pH 8.4.

$$TA\ (meq\ g^{-1}) = \frac{(V_b - V_s) \times N \times 10^3}{mg\ of\ sample}$$

V_b = Volume of standard acid used for blank

V_s = Volume of standard acid used for sample

N = Normality of the acid

10.2.2. Carboxyl Groups

The Ca acetate method

Extensive use has been made of the Ca acetate method for determining COOH groups:

$$2R.COOH + Ca(CH_3COO)_2 \rightarrow Ca(RCOO)_2 + 2CH_3COOH$$

The acetic acid liberated during the reaction is titrated with standard NaOH solution. The procedure (Schnitzer, 1972) is as follows: To between 50 and 100 mg of humic preparation in a 125-mL ground-glass stoppered Erlenmeyer flask, 10 mL of 1*N* $Ca(CH_3COO)_2$ and 40 mL of CO_2-free distilled water are added. A blank is set up simultaneously, consisting of 10 mL of 1*N* $Ca(CH_3COO)_2$ and 40 mL of CO_2-free distilled water only. After shaking for 24 hours at room temperature, the suspension is filtered and the residue is washed thoroughly with CO_2-free distilled water. The filtrate and the washings are combined and titrated potentiometrically with standard 0.1*N* NaOH to pH 9.8.

$$COOH\ (meq\ g^{-1}) = \frac{(V_s - V_b) \times N \times 10^3}{mg\ of\ sample}$$

V_b = Volume of standard base used for blank

V_s = Volume of standard base used for sample

N = Normality of the base

10.2.3. Phenolic OH [Acidic OH]

By Difference (Total acidity – COOH)

The quantity of phenolic OH (or more correctly, acidic OH) is computed as the difference between total acidity and COOH content.

Phenolic OH (meq g^{-1}) = Total acidity (meq g^{-1}) – COOH (meq g^{-1})

10.2.4. Carbonyl (C=O)

Formation of Oximes, Hydrazones and Carbazones

Most methods that have been used for determining C=O groups in humic substances are based on the formation of derivatives by reaction with such reagents as hydroxylamine, phenylhydrazone and 2,4-dinitrophenylhydrazine. The reactions can be followed in several ways including analysis of unused reagent and measurement of the increase in N content of the derivatives.

The procedure recommended by Schnitzer (1972) is as follows: To 50 mg of humic material in a 50-mL ground-glass stoppered Erlenmeyer flask, 5 mL of 0.25*M* 2-dimethyl aminoethanol solution plus 6.3 mL of 0.4*M* hydroxylammonium chloride solution are added. The system is heated on a steam bath for 15 minutes, cooled and the excess of hydroxylammonium chloride is back-titrated potentiometrically with standard perchloric acid solution. The end point is determined by plotting mV vs mL of acid. A blank is set up simultaneously consisting of 5 mL of 0.25*M* 2-dimethylaminoethanol solution plus 6.3 mL of 0.4*M* hydroxylammonium chloride solution only.

$$C = O\ (\text{meq g}^{-1}) = \frac{(V_b - V_s) \times N \times 10^3}{\text{mg of sample}}$$

V_b = Volume of standard acid used for blank

V_s = Volume of standard acid used for sample

N = Normality of the acid

Reference

Schnitzer, M. 1972. Chemical, spectroscopic and thermal methods for the classification and characterization of humic substances. In: Proc. Intern. Meetings on Humic Substances, Wageningen, pp. 293-310.

11

Adsorption of Organic Compounds by Clay

Understanding the mechanisms controlling the adsorption of organic molecules on clay minerals is of significance in many branches of science and industry. The interaction between clay minerals and organic compounds elicits interest that emanates from the prospect that the adsorption of the organic matter fraction in the soil on the mineral particles would render physical stability to soil aggregates.

Organic compounds tend to interact with clay minerals by:

- adsorption process at the external surfaces
- adsorption process at the external and internal surfaces
- exchanging exchangeable ions at the external surfaces
- exchanging exchangeable ions at the external and internal surfaces

11.1. Adsorption Isotherms

The adsorption of a solute from aqueous solution by a solid can be described by four basic types of adsorption isotherms (Fig.11.1). The L-type (for Langmuir) is the most common and describes the case where the solid has a high affinity for the solute. The S-type occurs when the solid has a high affinity for the solvent (e.g., water competes strongly with solute for adsorption sites). The C-type (constant partition) occurs when new sites become available as the solute is adsorbed. The H-type characteristically occurs when the solute has an unusually high affinity for the solid (such as chemisorption of positively charged organic species by the negatively charged clay).

Mathematical descriptions of adsorption isotherms (L-type) are carried out by the application of the Freundlich and Langmuir equations. The Freundlich adsorption equation is:

$$\mathrm{x/m} = KC^{1/n}$$

or, $\log (\mathrm{x/m}) = \log K + 1/n \log C$

where, x/m is the quantity of solute adsorbed per unit weight of adsorbent, C is the equilibrium concentration of the adsorbing compound and K and n are constants. In practical terms, a straight line is obtained when the data are plotted as log (x/m) vs log C. The intercept is equal to log K and the slope to $1/n$. The constant K provides an indication of the extent of adsorption.

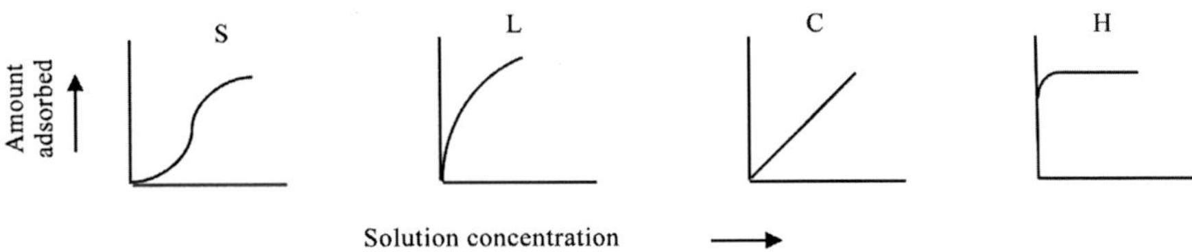

Fig. 11.1: Classification of adsorption isotherms

Theoretically speaking, the Freundlich equation represents a situation in which the quantity of solute adsorbed increases indefinitely with increasing concentration. This will happen when multilayers are formed.

The Langmuir equation, which was originally derived for the adsorption of gases by a solid, is as follows:

$$x / m = \frac{KbC}{1 + KC} \quad \text{or,}$$

$$C / x / m = \frac{1}{Kb} + \frac{C}{b}$$

where, x/m is the quantity of solute adsorbed per unit weight of adsorbent, C is the equilibrium concentration of the adsorbing compound and K is the constant related to the bonding energy and b is the adsorption maximum or total amount of solute capable of being adsorbed. In this case, a straight line is obtained when $C/x/m$ is plotted against the equilibrium concentration C. The Langmuir adsorption equation assumes a monolayer adsorption on a uniform surface with no interaction between adsorbed molecules.

11.2. Adsorption Mechanisms

Several adsorption mechanisms are involved in the adsorption of organic compounds by clay minerals, the main ones being: (1) Physical adsorption or van der Waals' forces, (2) electrostatic attraction or chemical adsorption, (3) hydrogen bonding and (4) coordination complexes. Two or more mechanisms may operate simultaneously, depending on the properties of organic species, nature of exchangeable cation on the clay, surface acidity and moisture content of the system.

11.2.1. Physical Bonding (van der Waals' Forces)

Physical or van der Waals' forces are rather weak and typically, these forces are a result of the fluctuations in the electric charge density of individual atoms. An electrically positive fluctuation in one atom tends to produce an electrically negative fluctuation in the neighbouring atom and results in a net attractive force. Attractive forces resulting from these fluctuations are assumed to exist between every pair of atoms or molecules.

A polar molecule has a permanent net asymmetry in the distribution of negative charges and molecules of opposite sign can line up through what is called 'dipole-dipole interaction'. The dipole moment of one molecule can produce a dipole in its neighbour, thereby leading to "induced" dipole interactions. The following shows the attraction of two molecules resulting from a line-up of positive and negative charges:

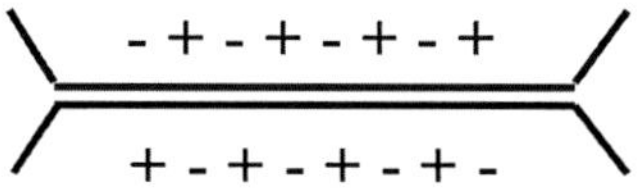

```
\   - + - + - + - +   /
 \___________________/
 /‾‾‾‾‾‾‾‾‾‾‾‾‾‾‾‾‾‾‾\
/   + - + - + - + -   \
```

Adsorption caused by physical forces can be of considerable importance in the adsorption of neutral, polar and non-polar molecules.

11.2.2. Electrostatic Bonding

Electrostatic bonding by clay and organic matter occurs through the process of cation exchange or protonation. Cation exchange takes place when positively charged organic cations replace inorganic cations on the exchange complex, as follows:

$$\boxed{\text{Clay}} - M^{+} + R\text{-}NH_3^{+} \rightleftarrows \boxed{\text{Clay}} - NH_3R + M^{+}$$

The adsorption by this mechanism is related to both the basic character of the organic molecule (chain length, type of cation, etc.) and the pH of the system.

Organic compounds that are weakly basic may be adsorbed at particle surfaces through protonation, a process by which an organic molecule assumes a positive charge by accepting an H^+ ion.

11.2.3. Hydrogen Bonding

This is a linkage between two electronegative atoms through bonding with a single H^+ ion. The H^+ ion is a bare nucleus with charge of +1 and has a strong tendency to share electrons with those atoms that contain an unshared electron pair, such as oxygen. The bond is weaker than ordinary ionic or covalent bonds, but stronger than van der Waals' forces of attraction.

Typically, H-bonding may occur at clay surfaces as follows:

R-O-H- - -O- [Clay]

R-CO-OH- - -O- [Clay]

A prime site for adsorption is surface oxygens or OH's; other possibilities include protons of adsorbed water molecules.

11.2.4. Coordination

The metal ion forms a bridge between the organic molecule and the soil constituent to which it is attached. A closely related bonding mechanism is a polyvalent cation acting as a salt bridge between the soil component and the COOH group of an organic molecule.

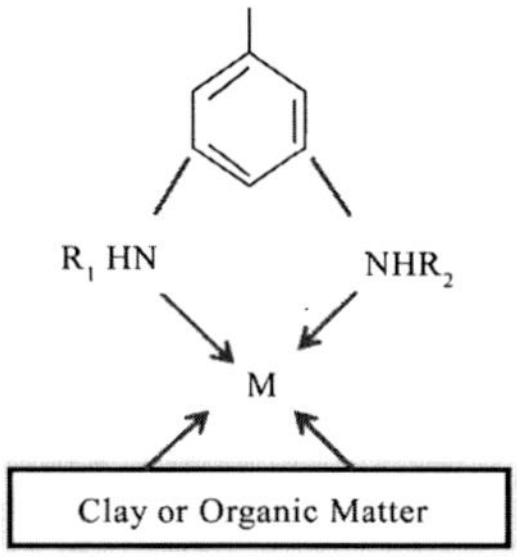

R_1 HN NHR$_2$

M

Clay or Organic Matter

[Clay or Organic Matter] - M - OOCR

11.3. Adsorption of Specific Classes of Compounds

Some points regarding adsorption of specific classes of compounds are given below:

Organic Acids: Negatively charged clay normally repels negatively charged organic species, with little or no adsorption. However, with the anionic character of organic acids being pH-dependent, some adsorption is possible through H-bonding and van der Waals' forces, particularly when the pH falls below the pKa of the acidic group (molecule in undissociated state).

Organic Cations, Amines and Amino acids: Clay colloids can tightly bind positively charged organic molecules through chemical adsorption. In case of amino acids, adsorption occurs through cation exchange below the isoelectric point of the amino acid.

Non-ionic Polar Molecules: A vast majority of uncharged organic molecules of polar character are adsorbed by clay minerals, mainly by H-bonding. Non-polar parts of the molecule may further be bound by the van der Waals' forces.

12

Clay-Organic Matter Complexes - I

Extensive studies have shown that much of the humified material in soil is finely bound to colloidal clay. However, it is uncertain to what proportion of the clay surface in any given soil is coated by organic substances as this will depend on the organic matter content and the type and amount of clay.

The ways in which organic substances are retained in soil are as follows:

1. As insoluble polymeric complexes of humic and fulvic acids.
2. As polymeric complexes of humic and fulvic acids bound together by di- and trivalent cations such as Ca^{2+}, Fe^{3+} and Al^{3+}.
3. As substances held on clay mineral surfaces. Organic matter may be bound through cation and anion exchange, bridging by polyvalent cations (clay-metal- humus), H-bonding, van der Waals' forces and in other ways.
4. As organic substances within the interlayers of expanding type clay minerals. This organic matter is not solubilized by conventional extractants, but is released by destruction of clay with hydrofluoric acid.

12.1. Proportion of Soil Organic Matter Bound to Clay

Differential extraction procedures have been used to study the state of combinations of organic constituents in soil. Tyurin's scheme as reported by Kononova (1966) is as follows:

Description	Type of linkages
Humic substances soluble in dilute alkalis without prior removal of exchangeable Ca.	Free forms of polymeric complexes of humic and fulvic acids.
Humic substances which are only soluble in dilute alkalis after the removal of exchangeable Ca with dilute mineral acids.	Polymeric complexes of humic and fulvic acids in linkages with Ca.
Humic substances soluble in dilute alkalis only after alternate acid and base treatments.	Complexes of humic and fulvic acids linked with relatively stable hydrated sesquioxides.

There were well-defined regularities in the way humic substances were combined in soils. In grassland soils, the humic substances were linked mostly with calcium; free compounds were virtually absent. On the other hand, in forest soils the humic substances were predominantly in the form of polymeric complexes readily solubilized by direct extraction with dilute alkali. The insoluble humic material in peat probably existed in polymeric forms held together by H- bonding. Results obtained by several workers for the proportion of the soil carbon contained in clay-organic complexes are given in Table 12.1. From 52 to 98% of the carbon in the soils was associated with clay.

Table 12.1: Proportion of soil organic carbon contained in the clay-organic complex

Soil	Total C in soil (%)	C in clay-organic Complex (% of total soil C)
Podzol	1.6	89.6
Chernozem	4.4	85.2
Rendzina	5.8	54.3
Brown earth	3.2	68.1
Red brown earth	2.2	71.5
Lateritic red earth	1.7	97.8
Solodized solonetz	1.0	76.4
Solonized brown soil	0.6	51.6

12.2. Interactions Involving Humic and Fulvic Acids

Humic and fulvic acids contain a variety of reactive functional groups that are capable of combining with clay minerals (Fig. 12.1) and the sorption of humic constituents by clay still provides an active *organic* surface for exchange with cations of the soil solution and for sorption of pesticides. The possible mechanisms of interaction between humic substances and clays have been summarized in Table 12.2.

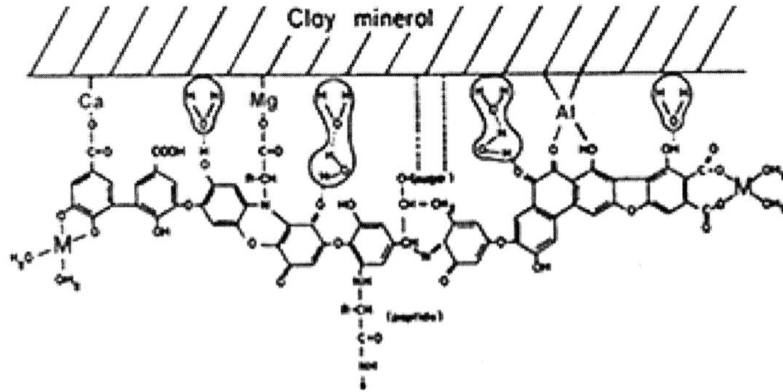

Fig. 12.1: Schematic diagram of a clay-humate complex in soil

Table 12.2: Possible mechanisms of bonding between humic substances and clay mineral surfaces

	Group on humic matter	Groups and sites on mineral	Remarks
A. Primary interactions			
Ion-dipole			By weakly hydrated and non-hydrated monovalent cations. Through polyvalent cations in dehydrated system. Desorbable by Na salts.
Cation bridging Water bridging	Carboxylate and uncharged polar groups such as NH_2, C = H, COOH and OH	Exchangeable cations at external basal surfaces. On inter- layer surfaces at low pH (4.5).	Polyvalent cations in hydrated systems. Desorbable by Na salts, water and ultrasonic treatment.
Anion exchange	Carboxylate	Crystal edge surfaces. Poly-hydroxy compounds of Fe and Al at external, basal surfaces	Under acidic pH conditions and when exchange sites are occupied by Fe and Al ions. Desorbable by other anions in solution as Cl^- and NO_3^-.
Ligand exchange	Carboxylate	Crystal edge surfaces. Poly-hydroxy compounds of Fe and Al at external, basal surfaces	When exchange sites are occupied by Fe and Al ions. On hydrous oxides of Fe and Al. Desorbable by other more strongly adsorbed anions such as hydroxyl.
B. Secondary interactions			
Hydrogen bonding	Amino, COOH, OH	Oxygens of external, basal surfaces	Weak
van der Waals'	Uncharged portions of polymer chain	Mainly at external, basal surfaces	Short-range, weak, but additive.
C. Indirect interactions			
Entropy effects			Due to entropy gain caused by displacement of water molecules from clay surface by a single polymer chain.

12.2.1. With Clay Minerals Having Mica-Type Lattices (Montmorillonites, Illites and Vermiculites)

Organic anions are normally repelled from negatively charged clay surfaces; therefore, clay minerals such as montmorillonite adsorb humic and fulvic acids only when polyvalent cations are present on the exchange complex. Unlike monovalent Na^+ and K^+ ions, polyvalent cations are capable of maintaining neutrality at the surface by neutralizing both the charge on the clay and the acid functional group of organic matter [e.g., COO^-].

The main cations binding humic and fulvic acids to soil clays are Ca^{2+}, Fe^{3+} and Al^{3+}. Of these, the divalent Ca^{2+} ions do not form strong coordination complexes with organic molecules and would be effective only to the extent that a bridge linkage could be formed. Organic matter bound thus would rather be easily displaced by a monovalent cation, which may account for the small quantities of humic materials which can be displaced when Ca- saturated soils are leached with an NH_4^+ salt.

In contrast, Fe^{3+} and Al^{3+} form coordination complexes with organic compounds and strong bonding of humic substances are possible through this mechanism. In this case, displacement of bound metal is difficult and may require a strong chelating agent.

Thus, in the adsorption of organic polyanions by mica-type clay minerals two major types of interactions may be involved. In the first type of interaction, the polyvalent cation acts as a *bridge* between two charged sites. In this case hydration water of the exchange cation is not displaced, but forms a H-bond with the organic anion. In the second type, the organic anion becomes coordinated to the cation, with the displacement of a water molecule from the hydration shell. For a long chain organic molecule, several points of attachment to the clay particles are possible.

Other bonding forces that operate between organic anions and the clay surface include H- bonding between polar groups of the organic molecule and adsorbed water molecules or oxygens of the silicate surface. The strength of an individual bond is small, but additive.

Stringent drying (such as desiccation or consumption of available moisture by plant roots) increases the bonding between humic material and clay by eliminating water of hydration and bringing the humic matter in close contact with the clay.

12.2.2. With Hydrous Oxides

When clay minerals are coated with layers of hydrous oxides, it is these hydrous oxides that dominate the surface reactions more than the clay. Chances for strong bonding exist through ligand exchange, as well as through simple anion exchange. Anion exchange is a distinct possibility due to some positive sites normally existing on iron and aluminium oxides at $pH < 8$. By simple Coulombic attraction, the oxides can be associated with organic anions.

Coordination or ligand exchange takes place when the anionic group pervades the coordination shell of Al or Fe and is incorporated with the surface OH layer. The sorption of fulvic acid on oxide surfaces accompanies displacement of OH groups by COO^- ions, indicating ligand exchange. Contrary to anion exchange, the organic anion is not easily displaced with simple salts, even though adsorption is pH sensitive. As in the case of organic cations on the surface of clay minerals, a very strong bond will result if more than one group on the humic molecule participate.

Humic materials may interact with crystalline oxides in an identical way. In soils developed from volcanic ash, allophanes are strong adsorbents of humic substances, which is why there are exceptionally high levels of organic matter in soils with high allophane content.

13

Clay-Organic Matter Complexes - II Role of Organic Matter in Forming Stable Soil Aggregates

Organic matter is of immense importance in forming good aggregates in a wide range of soil types, particularly those representing Mollisols, Alfisols, Ultisols and Inceptisols. Organic matter is somewhat less important in the Oxisols where hydrous oxides may play apredominant role. These relationships are depicted in Fig. 13.1.

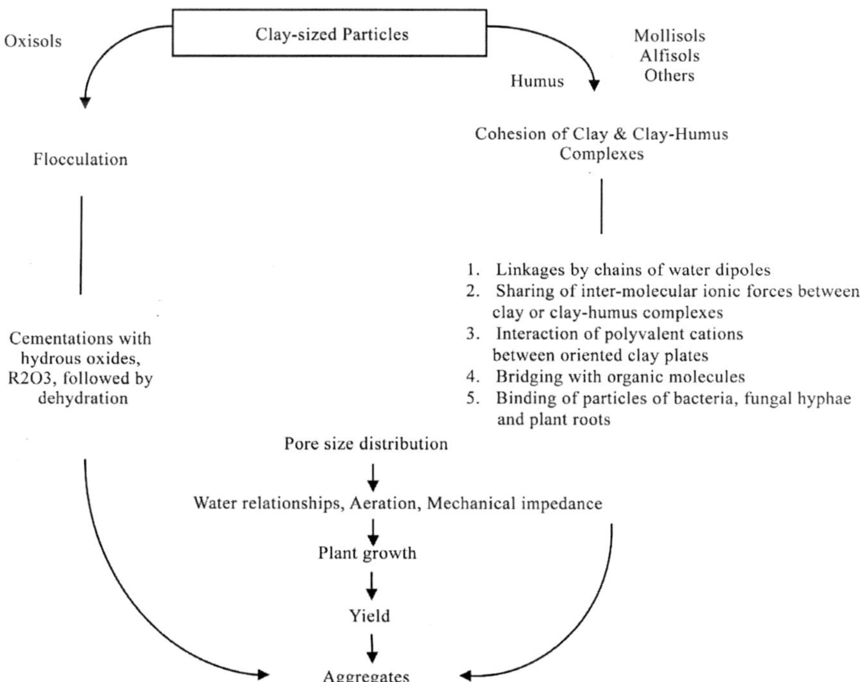

Fig. 13.1: Relationship between OM and the formation of soil aggregates

Aggregates do not directly influence plant growth, but they alter the physical and chemical environments in which plant roots grow through their effects on porosity, aeration, moisture retention, etc., when well-irrigated, even fine-textured soils permit adequate exchange of gases with the atmosphere. The polysaccharides of soil have received the greatest attention in producing stable aggregates, but other organic substances bonded to clay through association with Fe or Al may also be involved.

13.1. Specific Action of Organic Substances

Soil aggregation may be influenced by organic constituents in at least three different ways. First, organic substances serve as binding agents for the cohesion of clay particles, such as through H-bonding and coordination with polyvalent cations. Flocculation of clay is an obligate pre-requisite for aggregation. A wide variety of organic compounds may be involved in the cementation of clay or clay particles, including humic and fulvic acids which are linked to clay as clay-metal-humus complex. Clay particles themselves may cohere and bridge between larger sand and soil grains. However, if devoid of organic coatings, clay is easily dispersed by the slaking action of water.

Second, gelatinous organic materials surround soil particles and hold them together through a cementing action. Whereas a variety of compounds may be responsible, a major role is played by polysaccharides. Living bacteria and other microbes have been shown in laboratory studies to bind soil particles together and this has been attributed to the polysaccharides they secrete. Since polysaccharides, for most part, are readily attacked by microbes, their effect on aggregation is short-lived.

Third, soil particles are held together through physical entanglement by fungal hyphae and microscopic plant roots. The mycelia and fungi are often extended throughout the soil and particles are entrapped and tied together. This is sometimes apparent with naked eye; under magnification, small clumps of soil particles can often be seen clinging to the mycelia. Aggregation through this mechanism is ephemeral and is dependent on sustaining a high content of fungal hyphae. The fibrous root systems of the grasses ramify and open up the soil; they also encompass individual crumbs in a net-like web to form clusters defiant to the slaking action of water.

The synthesis of aggregates is influenced by (1) the kind and amount of organic matter in the soil, especially gums and mucilages, (2) the presence or absence of fungal hyphae and microscopic plant roots, (3) wetting and drying, (4) freezing and thawing, (5) nature of cation on the exchange site and (6) the burrowing action of soil animals, especially earthworms.

13.2. Clay Domain Theory of Aggregate Formation

According to Emerson (1959) theory of soil aggregate formation, crumbs are formed from units of colloidal clay, or domains and coarser particles of silt and sand cemented together by humus. A domain is defined as "a group of clay particles with suitable exchange cations which are oriented and sufficiently close together for the group to behave in water as a single entity." The possibilities for four types of bonds are: (1) Quartz-organic matter-quartz, (2) quartz-organic matter domain, (3) domain-organic matter-domain (organic matter positioned between the faces of two clay domains, between two edges and between an edge and a face), and (4) domain-domain, edge-face.

While the theory explains many of the properties of soil crumbs, it does not take into account the extensive occurrence of clay-organic complexes. In soils that are well supplied with organic matter and which are well aggregated, most of the clay will be coated with organic matter.

13.3. Significance of Organo-mineral Aggregates

Organo-mineral associations evolve as the fundamental building units of (micro-) aggregates in soil during pedogenesis. Presumably, these organo-mineral associations which are composed of soil minerals and organic matter of various origins serve as the key for nutrient provision, preservation and carbon storage. The dominant inorganic constituents in such aggregates are secondary soil minerals (e.g., metal (hydr-)oxides, clay minerals) and persistent primary minerals (e.g., quartz).

The organic matter in soil is a product of microbial and root bio- and necromass, biogenic excretion and the decomposition of floral and faunal tissues and debris. Surface interactions between minerals and organic matter are assumed to influence the formation of organo-mineral associations and aggregates. Composite building units, representing the smallest size class of microaggregates, emerge if at least two of these (micro-) aggregate forming materials are combined. Consequently, organic matter incrusted by minerals or organic matter-coated minerals form as the smallest (size < 2 μm) and most stable building units of aggregates.

The stability of aggregates against external forces (e.g., wetting and osmotic stress) depends on their intra-aggregate bonding which are mediated by organic and inorganic binding agents, enabling aggregation by their attractive surface properties, agglutinating and cementing characteristics. Consequently, the stability of aggregates against external forces differs, for instance, with organic carbon and clay content.

Adsorption of organic matter on and co-precipitation of organic matter with soil minerals also protect organic matter against decomposition. Several essential soil functions, such as habitat provision, nutrient storage, and water retention, are related to the presence and arrangement of these organo-mineral associations in a superordinated three-dimensional aggregate structure.

14

Humus – Pesticides Interactions in Soil

It is well known that organic matter plays a unique role in the binding of pesticides in soil, and that this phenomenon is usually the most important cause for interaction of pesticides in the soil environment. Fulvic or humic acids are most commonly associated with the binding interactions. Binding can take place with either the original pesticide or a transformation product, the reaction being induced by abiotic or biotic agents (microbial or plant enzymes). The reactions or processes involved are apparently the same as those responsible for the formation of humic substances, i.e., for the humification process.

Binding of pesticides to organic matter can occur by sorption (van der Waals' forces, hydrogen bonding, hydrophobic bonding), electrostatic interactions (charge transfer, ion exchange or ligand exchange), covalent bonding or combinations of these reactions. The formation of covalent linkages between pesticides and humus constituents and fulvic or humic acids in the presence of phenol oxidases or clay minerals has been demonstrated. With chlorinated phenols and carboxylic acids, it is practically feasible to isolate and identify cross-coupling products and to decode the site and type of binding. The binding of chlorinated phenols to humic substances has been demonstrated by using ^{14}C-labelled chemicals and by measuring the uptake of radioactivity by the humic material.

Three major factors determine the extent to which pesticides are adsorbed by soil organic matter: (1) Physical and chemical characteristics of the adsorbents (chiefly, humic colloids), (2) nature of the pesticide and (3) properties of the soil system, such as the clay mineral composition, pH, kinds and amounts of exchangeable cations, moisture and temperature.

14.1. Organic Matter versus Clay as Adsorbent

Organic matter and clay are the two soil components that are more often than not entangled in pesticide adsorption. However, individual effects are not easily ascertained for the reason that, in most soils, organic matter is intimately bound to clay, probably as clay-metal-organic complex. Two different types of adsorbing surfaces are thus normally available to the pesticide, *viz.,* clay-humus and clay alone. Accordingly, clay and organic matter function more as a unit than as separate entities and the relative contribution of organic and inorganic surfaces to adsorption will depend on the degree to which the clay is coated with organic substances.

Interaction of organic matter with clay furnishes an organic surface for adsorption. Up to an organic matter content of about 8%, both organic and mineral surfaces are involved in adsorption; at higher organic matter contents adsorption occurs mostly on organic surfaces. It should be noted that the amount of organic matter required to coat the clay will vary from one soil to another and will depend on both the kind and amount of clay in the soil. For soils having similar clay and organic matter contents the contribution of organic matter will be the highest when the predominant clay mineral is kaolinite and lowest when the predominant clay mineral is montmorillonite. The adsorption capacity of clays for pesticides follows the order: montmorillonite > illite > kaolinite.

A given quantity of soil is added to a pesticide solution of known concentration; the mixture is allowed to equilibrate and the concentration of the pesticide in the solution phase is estimated. Subsequently, the quantity of pesticide adsorbed is calculated from the decrease in concentration and is normally expressed as μmoles adsorbed per kg of soil (x/m). By repeating the measurements at several pesticide concentrations, an adsorption isotherm can be obtained by plotting the quantity adsorbed (x/m) vs. the equilibrium concentration (C). In most instances, a straight line is obtained when the data are plotted as log (x/m) vs. log C, according to Freundlich adsorption equation $x/m = KC^{1/n}$ where K and n are constants.

A modified version of the Freundlich equation has often been used to express adsorption data. When n is near unity, the constant K (referred to as distribution coefficient K_d) is given by

$$K_d = \frac{\text{Pesticide adsorbed } (\mu\text{moles kg}^{-1} \text{ of soil})}{\text{Pesticide in solution } (\mu\text{moles litre}^{-1})}$$

14.2. Qualitative Differences in Soil Organic Matter

While great differences exist among soils in regard to their organic matter contents, major qualitative differences also do exist, not only with respect to the known classes of organic compounds (lipids, carbohydrates, proteins) but also with the humic substances (humic acid, fulvic acid, etc.). For example, the percentage of organic matter as fats, waxes and resins ranges from as little as 2% in some soils to over 20% in certain others. The percentage of organic matter as "protein" may vary from 15 to 45% and the carbohydrates range from 5 to 25%. Humified organic matter may comprise three-fourths of the total organic matter in some soils, but less than one-third in some others. The humic material of grassland soils is dominated by humic acids, while that in forest soils is relatively rich in fulvic acids.

Because of their low molecular weights and high activities, fulvic acids are more soluble than humic acids and they may have special functions in respect of pesticide transformations. First, they may act as transporting agents for certain pesticides in soils and natural water. Second, fulvic acids, by virtue of their functional group content, may catalyze the chemical decomposition of certain pesticides.

14.3. Chemical Reactions Involving Pesticides and Organic Substances

Organic fraction of the soil has the potential for promoting the non-biological degradation of many pesticides. Organic compounds containing reactive groups such as COOH, phenolic-, enolic-, heterocyclic- and aliphatic-OH and others produce chemical changes in a variety of pesticides. Humic substances are strong reducing agents and have the capability of bringing about a variety of reductions and associated actions. The occurrence of stable free radicals in humic and fulvic acids further implicates organic matter in chemical transformations of pesticides.

14.4. Chemical Binding of Pesticides and Their Decomposition Products

Pesticide-derived residues can form stable chemical linkages with components of soil organic matter and such binding greatly increases persistence of the pesticide residues. Two main mechanisms are involved: (1) direct chemical attachment of the residues to reactive sites on colloidal organic surfaces and (2) incorporation into the structures of newly formed humic and fulvic acids during the humification process.

Another factor to consider is the partial degradation of many pesticides by microbes leading to the formation of chemically reactive intermediates which can combine with amino- or carbonyl-containing compounds. Thus, loss of the side chain from the phenoxy alkanoic acids by enzymatic action leads to the formation of phenolic constituents which can either be oxidised further *via* the enzymatic route or undergo condensation (may be, as quinines) with amino compounds to form "humic-like" substances. On the other hand, amines produced by biological decomposition of pesticides may react with carbonyl constituents of organic matter. By virtue of being highly reactive, decomposition products of some of the pesticides can become integral part of the pool of precursor molecules for humus synthesis and thus, lose their identity.

14.5. Adsorption Mechanisms

Binding of pesticides to organic matter can occur *via* sorption (van der Waals' forces, hydrogen bonding, hydrophobic bonding), electrostatic interactions (charge transfer, ion exchange or ligand exchange), covalent bonding or a combination of these reactions. [Please refer to § 11.2 for details.]

14.5.1. Ion Exchange and Protonation

Like the secondary silicate minerals [montmorillonite, illite, kaolinite], soil organic colloids are negatively charged, although positive sites may be present under some conditions through free amino groups. Adsorption of pesticides by this mechanism is largely restricted to those types that exist as cations [as RNH_3^+] or which can become positively charged through protonation.

$$[OM]\text{-}COO^- + RNH_3^+ \rightarrow [OM]\text{-}COO.H_3NR$$

Anion exchange may be possible in some cases through interaction of anionic pesticides to positively charged spots on organic surfaces. Other possible combining sites on organic matter include a COO^- ion plus a phenolate ion combination and a COO^- (or phenolate ion) plus a free radical site.

Factors that influence the availability of exchange sites for adsorption include the presence of competing metal cations and pH. Soil pH has a direct effect on the relative importance of organic matter and clay in retaining organic cations. Unlike clay, organic colloids have a strongly pH-dependent charge. Therefore, the contribution of organic matter to CEC, and subsequently to retention, will be higher in neutral and slightly alkaline soils than in acid soils. For each unit change in pH, the change in CEC for organic matter is several times greater than for clay. In a typical temperate zone soil, the contribution of organic matter to the CEC may range from about 50% at pH 7.0 to only 25% at pH 5.0.

Less basic pesticides may become cationic through protonation. Whether or not protonation occurs will depend upon: (1) the nature of the pesticide and (2) the proton supplying power of the humic colloids. Reactions leading to the adsorption are shown by the following equations:

1. $T + H_2O \leftrightarrow HT^+ + OH^-$
2. $RCOOH + H_2O \leftrightarrow RCOO^- + H_3O^+$
3. $RCOO^- + HT^+ \leftrightarrow R.COO.HT$
4. $RCOOH + T \leftrightarrow R.COO.HT$

Where R is the organic colloid; T is the pesticide molecule, HT^+ the protonated molecule and H_3O^+ the hydronium ion.

Equation 1 represents pH-dependent adsorption through protonation in soil solution.

Equation 2 represents ionization of the colloid COOH group.

Equation 3 shows ionic adsorption of the cationic pesticide formed by reaction 1.

Equation 4 shows adsorption through direct protonation on the surface of the organic colloid.

For anionic pesticides repulsion by the predominantly negatively charged surface of organic colloids may occur. Positive adsorption of anionic pesticides at pH values below their pKa values can be ascribed to the adsorption of the *unionized* form of the pesticide to organic surfaces, such as by H-bonding between COOH group and C = O or NH groups of organic matter.

$$R - \overset{\overset{\displaystyle O}{\|}}{C} - OH \text{ ---- } O = C - \boxed{\text{Organic Matter}}$$

The metal ion forges a bridge between the soil constituent to which it is attached and the organic molecule. A closely related bonding mechanism is where a polyvalent cation functions as a salt bridge between the soil component and the COOH group of an organic molecule.

14.5.2. Hydrophobic Bonding

Partitioning on hydrophobic surfaces is a mechanism for retention of non-polar organic compounds by soil organic matter. Essentially, hydrophobic adsorption results from a weak solute-solvent interaction; that is, the low solubility or hydrophobic nature of the solute. Adsorption by this means is referred to as the “squeezing-out” of the molecule from solution and its accumulation at the solid interphase where competition with the solvent is minimum.

Hydrophobic adsorption increases as the solute becomes more and more non-polar or as water solubility decreases.

Active surfaces for hydrophobic interaction include fats, waxes and resins, as well as possible aliphatic side chains on humic and fulvic acids. Humus, by virtue of its aromatic framework and presence of polar groups, may contain both hydrophobic and hydrophilic adsorption sites.

15

Environmental Significance of Humic Substances

Humic substances are the most widely-present, natural, complexing ligands occurring in nature. They are found not only in soils, but in natural waters, sewage, compost heaps, marine and lake sediments, peat bogs, carbonaceous shales, coals and other miscellaneous deposits.

The fate of natural organic matter, especially humic substances, has over recent decades attracted increasing attention of scientists representing various disciplines. Humic substances constitute about 25% of the total organic carbon on the earth. These are a group of naturally occurring complex molecular structures into which biomolecules derived from plant and animal residues are transformed progressively through biotic and abiotic pathways, by aggregation and accumulation processes.

Research on humic substances is complex since these organic compounds are associated with soil mineral fractions and require physical and/or chemical separation from the inorganic components by an extraction procedure, prior to their physico-chemical analysis. The most efficient of these separation procedures involves an extraction with alkali which operatively identifies three humic substances fractions, based on their water solubility, *viz*., fulvic acid which is soluble at all pH values, humic acid which is not soluble under acidic conditions (pH <2) but soluble in solution of higher pH values and humin which is not soluble in water at any pH value.

Humic research concerns investigations of products obtained as a result of their extraction. Some research technologies applied to humic substances require extracted and purified material. Functions and reactions of humic substances do not necessarily require extraction. Some studies applying ^{13}C-NMR spectra compared the spectra of extracted material with those of original soil material. Both sets of spectra were found to be the same, thereby indicating that the extraction does not change the structure of the humic substances. Alkaline extraction of soil and natural waters produces organic matter fractions that have great ecological relevance and are involved in environmental processes.

Thus, humic substances research is the only option to explore and elaborate the important and complex role of these substances in the ecosystem.

Humic substances are important in geochemistry for the following reasons: (1) They may be involved in the transportation and subsequent concentration of mineral substances, such as ores and nodules of marine strata; (2) they may be responsible for the enrichment and concentration of uranium and other metals in various bioliths, including coal.

15.1. Humic Substances and Terrestrial Ecosystems

Undoubtedly, a driving force for the growing interest in humic substances is the increasing awareness in their immense role in terrestrial and aquatic systems, as well as their bio-stimulatory effects on plant growth. Soils contain more organic carbon than the atmosphere and vegetation combined; therefore, mineralization of soil organic matter and release of carbon versus humification processes could greatly affect the global climate change.

Humic substances are known to be a major component of soils and they are crucial to human welfare due to their role in generating food, fodder and fibre. They closely relate to the biogeochemical cycling of nutrients and anthropogenic organics as well as some polluting elements. As a consequence, humic substances are able to affect crops' contaminant uptakes and/or leaching into groundwater. Studying the interactions of heavy metals or organic compounds with humic substances provides insights into pollutant mobility, behaviour, and fate and allows better understanding of natural mitigation phenomena and potential impact on the environment.

The ability of the soil to function as a "sink" for sulphur and nitrogen oxides in the atmosphere may be partly because of reactions involving organic colloids. Humus strongly influences the sorption capacity of the soil for a variety of organic and inorganic gases.

Due to the role of humic substances in many complex chemical and biochemical reactions in soils, humic substances largely determine soil quality and they are vital to maintaining soil fertility. The ability of fulvic acid and humic acid to dissociate H^+ ions from various functional groups (mostly carboxyl and phenolic) results in an array of different negative charges which markedly contribute to the soil cation exchange capacity. The sorption capacity of humic substances, especially fulvic acid, exceeds by far (up to a factor of ten) the sorption capacity of soil mineral components and often dominates their sorption properties, especially in sandy or sandy loam soils. Humic substances also have a decisive impact on the pH buffering capacity of soils. Furthermore, they affect physical, chemical and biological properties of soils to a much greater extent than other soil mass constituents.

An increase in humic substances content, especially that of humic acid in the presence of Ca, causes the formation of stable aggregates. This, in turn, positively affects field water capacity, air capacity, porosity and permeability of different soil types. Important losses of soil organic matter, especially humic acid, contribute to an increase in soil compactness (mainly in sandy and sandy loam soils), a decrease in soil aeration and consequent reductive chemical conditions.

Among others, the dark colour of humic substances enhances sunlight absorption and improvement of the thermal properties of soils which is crucial, especially at the beginning of the growing season. They are also the main source of energy for decomposition of organic matter. Finally, humic substances are also formed during aerobic oxidation of different waste materials (composting), potentially recycled into soil amendments and are beneficial for soil fertility and quality.

15.2. Humic Substances and Aquatic Ecosystems

Soil organic matter is a major source of organic carbon in aquatic systems. Thus, humic substances are extensively investigated by the community of aquatic chemists. Aquatic carbon is important as a part of the global carbon cycle, and also for local biogeochemical processes in lakes, streams and rivers.

Humic substances in water directly and indirectly affect aquatic ecosystems, as well as living organisms. In the long term, dissolved humic substances affect primarily the physical and chemical properties of waters and can act as the most important purifying natural components in these environments, e.g., by stimulating biotransformation of xenobiotics. In the short term, they can act as a source of organic nutrients and also exhibit the capacity to regulate the bio-concentration and toxicity of xenobiotics and heavy metals.

Dissolved humic substances affect the optical properties of natural waters. They directly impact both the quantity and quality of light in waters. The strong specific UV absorption renders humic substances like a natural biogeochemical filter against specific UV radiation. An important issue is the chemical removal of pollutants by humic substances in water treatment and purification technologies. On the other hand, humic acids also act as contaminants due to their reactions with disinfectants which can generate by-products that are toxic to humans.

It has been found that humic substances can enhance biotic and abiotic degradation of phenols, polyaromatic hydrocarbons and pesticides in the aquatic environment. They are often deemed to be responsible for the binding of most of the available metal ions in water and soil. Interaction between humic substances and hazardous chemical wastes is most probable to occur when such wastes have been disposed off underground.

15.3. Humic Substances and Waste Waters

Waste waters treated by biological treatment processes contain significant amounts of organic matter. Humic substances and other organics are considered undesirable because they may react with chlorine (during chlorination) to produce carcinogenic chloroform and other halogenated organics, some of which are known to be toxic. Also, they may concentrate compounds such as pesticides and heavy metals, as well as interfere with removal of odour-producing compounds by competing for adsorption sites during activated carbon treatment. A major problem in the treatment of wastewaters for reuse is removal of humic substances and other organics.

The main task of humic substances in the environment is to remove toxic metals, anthropogenic organic chemicals and other pollutants from water. Ion exchange materials based on Ca humate have been found to be suitable for the removal of metals like Fe, Ni, Hg, Cd and Cu from water and to remove radioactive nuclides from water discharged from nuclear power plants. Their selective binding capabilities are also utilized for the annihilation of munitions and chemical warfare agents.

Humus-based filters have been advanced for sewage purification, with multiple applications. The filters are useful to clean chromate smelter waste water, to remove oil and dyes from waste waters and aquatic systems, to filter urban and industrial waste waters, to eliminate pesticides from sewage and phenols from water. Humus-containing materials have been also utilized for sorbing gases, e.g., the removal of waste gases from an animal carcass rendering plant.

15.4. Humic Substances and Environmental Remediation

High adsorption capacity, high ion exchange capacity and environmental compatibility make humic substances an attractive material for environmental remediation. Treatability studies on the application of humates in environmental remediation indicated that humates can detoxify organic and inorganic inhibitors of biological processes. Humates also enhance biodegradation of toxic organic substances (phenols, formaldehyde, mineral oil), thus making their treatment more efficient. Chemical studies have demonstrated that humates can be successfully used for immobilization of heavy metals (Cu, Fe, Mn). Thus, humates have the potential to be employed as a filling material for barrier walls to inhibit transport and bioavailability of heavy metals in soil.

Humic substances form water-soluble complexes with many metals, including radionuclides. These organics may, therefore, be important as radionuclide transport agents through the environment. Humic substances present in natural waters can influence the uptake of radionuclides by natural solids and their migration to surface and ground waters.

Humic substances, because of their ability to absorb organic pollutants from the environment, have been observed to be useful to remove contaminants such as herbicides, fungicides, insecticides, nematicides and some pharmaceutical products from water. The complex nature of the interaction between humic substances and xenobiotics and their influence in the environmental quality (water, soil and atmosphere) has been studied by different authors.

15.5. Humic Substances and Soil Ecology

The interactions between humic substances and microorganisms have been intensively studied for several decades. It has been reported that fermenting bacteria could reduce humic substances. This fact has significant implications for the ecology of anaerobic bacteria in soils and sediments. The cumulative production of acetate during this process is energetically advantageous for fermenting bacteria.

Several researchers have highlighted the role of soil organisms in soil structuring. Bacteria, fungi (and especially earthworms) shape soil structure. Additionally, they actively release organic exudates, including root mucilage, earthworm mucus and microbial extracellular polymeric substances in the surrounding soil. Hence, soil biota alters the environmental conditions and physicochemical properties (e.g., pH, redox-potential, electrical conductivity, density, availability of nutrients) by soil processing, growth and metabolism. These activities lead to a close relationship between aggregate and organic matter dynamics, especially in biological processes such as bioturbation.

The resultant biogenic structures and aggregates present altered colour, composition, increased water stability, hydrophobicity and carbon content as compared with the bulk soil. Consequently, *biogenic aggregates,* which can be physically and chemically distinct from physicogenic aggregates, are defined as formed under the influence of organic matter, roots, and microbial exudates, in addition to the compressive stress of roots and hyphae and soil fauna activity.

However, physical, chemical and biogenic aggregation processes do not run isolated within the distinct spatial and functional domains in soil. Interdependencies between these mechanisms rather establish dynamic equilibria. "Biogenic aggregation" is thus defined as the sum of all processes exerted or mediated by soil biota that either trigger or alter physical processes (e.g., compaction by mechanical stress due to growing roots) and chemical aggregation processes (e.g., by excretion of organic matter that acts as aggregation agent).

16

Characterization of Soil Organic Matter by UV-Visible Spectroscopy

Humic substances, the natural organic materials, represent a mixture of relatively small organic components, which form supramolecular structures held together by dispersive forces such as π - π and van der Waals' interactions. Identification of the best analytical method for complete characterization of humic substances is still being discussed. Humic substances differ in molecular weight, elemental composition, acidity and cation exchange capacity and often are classified into three major fractions according to their solubility as humic acids, fulvic acids and humins.

The humic acid fraction consists of hydroxyphenols, hydroxybenzoic acids and other aromatic structures with linked peptides, amino compounds and fatty acids. Fulvic acids are typically composed of a variety of phenolic and benzene carboxylic acids. Fulvic acid molecules are held together by hydrogen bonds to form stable polymeric structures or by association with polysaccharides that are easily separated by adsorption on charcoal or by gel chromatography. Fulvic acids contain more oxygen, less carbon and more acidic functional groups, particularly –COOH, as compared to humic acids.

Humins are insoluble fractions of the humic type polymers that form strong associations with minerals. A difficulty with the chemical extraction of humic substances is that they are tedious, labour intensive and are not suitable for large numbers of samples. New approaches in spectrometry that include an extensive array of the spectroscopic techniques have been successfully applied.

16.1. Spectroscopic Methods in Soil Organic Matter Studies

Spectroscopic measurements in the different regions of the electromagnetic spectrum have wide applications to the study of soil organic matter, especially, humic substances. Attractive features of most of these methods are: (1) they are non-destructive, (2) only small sample weights are required and (3) most are experimentally simple and do not require manipulative skills.

The spectroscopic methods that have been used in studies on soil organic matter are:

- Ultra-violet visible absorption spectroscopy
- Fluorescence spectroscopy
- Infrared spectroscopy
- Raman spectroscopy
- Electron spin resonance spectroscopy
- Mossbauer spectroscopy
- Nuclear magnetic resonance spectroscopy

16.2. Ultra-violet and Visible Regions

Absorption in the UV (200 - 400 nm) and visible (400 - 800 nm) regions is caused by atomic and electrometric variations and involves elevation of electrons in σ-, π- and η-orbitals from the ground state to higher energy levels. Constituents containing unbonded electrons on oxygen and sulphur atoms are capable of showing absorption, as well as systems containing conjugated C = C double bonds.

Absorption spectra of humic and fulvic acids in both the UV and visible regions are somewhat featureless; that is, well-defined maxima and minima are absent. However, a slight maximum is often indicated in the 260 - 300 nm region of the ultra-violet. Considerable variation exists in the optical density of humic substances from various sources.

Agricultural applications of humic substances are well known. Their use as heavy metal absorbents and for pollutant sequestration is documented. Visible spectral lines and indices ($Q_{4/6}$) are able to quite well characterize the quality, maturity and condensation degree of humic substances. It has been demonstrated that absorbance at wavelength 465 nm is equal to light absorption of components associated with the first phases of humification process (young humic substances). Light absorption at 665 nm pertains to well humified components. Low $Q_{4/6}$ values (< 4) indicate high quality of humic substances, which is known by Chernozems, usually.

The composition of organic matter is frequently characterized by determining the ratio of absorbances at 465 (E_4) and 665 (E_6) nanometers. The absorbance at 465 nm is due to smaller molecules and at 665 nm is due to larger molecules. The E_4/E_6 ratio is expected to be greater for larger molecular weight humic acids and smaller for smaller molecular weight fulvic acids. The E_4/E_6 ratio is < 6 for humic acid and 6 - 18.5 for fulvic acid. This ratio has been reported to be independent of concentrations of humic materials but to vary for humic materials extracted from different soil types.

17

Characterization of Soil Organic Matter by Nuclear Magnetic Resonance Spectroscopy

The nuclear magnetic resonance spectroscopy (NMR) has led to major advances in the studies on the nature and chemical composition of soil organic matter. The technique has the potential for characterizing organic matter in the intact soil and its size fractions without the need for extraction and fractionation. Furthermore, it can provide information on the compositional changes in crop residues, peat and litter of the forest soils during biodegradation and humification.

17.1. Theory of NMR Spectroscopy

Nuclear magnetic resonance spectroscopy is a form of absorption spectroscopy. Akin to other spectroscopic methods, NMR spectroscopy depends upon the interaction of electromagnetic radiation with nuclear, atomic or molecular species.

All nuclei carry a charge. In some nuclei (as ^{13}C and ^{1}H), this charge spins about the nuclear axis and thereby generates a magnetic dipole along the axis. Under the influence of an external magnetic field, H_0, the nuclei precess about the magnetic field. The angular momentum of the spinning charge is described in terms of spin number, I, which specifies the number of orientations that a nucleus may assume in an external magnetic field, in accordance with the formula $2I + 1$. For both ^{13}C and ^{1}H, $I = ½$; thus, each can have two energy levels.

When an electromagnetic radiation is applied to generate a small alternating field, H_1, at right angles to the external magnetic field, H_0, the molecule absorbs energy. At the point where the frequency matches the precession frequency, the nuclei flip over or resonate, thereby inducing a voltage change (resonance signal), which is amplified and recorded.

The fundamental equation relating the frequency of electromagnetic radiation, v, to the magnetic field strength, H_0, is given by:

$$v = \frac{\gamma H_0}{2\pi}$$

where γ, the gyromagnetic ratio, is constant for particular nuclear group. Of the elements useful in soil organic matter studies, ^{1}H has the most favourable gyromagnetic ratio, followed, in order, by ^{31}P, ^{13}C and ^{15}N. With ^{1}H assigned the value of unity, the sensivities for the four elements are 1, 0.405, 0.251 and 0.092, respectively.

The value obtained for v is dependent on the strength of the magnetic field. To facilitate comparisons, results of an NMR experiment are expressed in terms of "chemical shifts" with reference to a reference standard. The chemical shift is given by:

$$\delta = \frac{v_{sample} - v_{reference}}{v_{reference}} \times 10^6$$

For both ^{13}C- and ^{1}H-NMR spectroscopy, the most common reference standard is an aqueous solution of tetramethylsilane, $Si(CH_3)_4$.

In NMR spectroscopy, the experimental sample is placed in a uniform magnetic field and a second magnetic field is introduced. Either the magnetic field or the frequency of the oscillaing field is varied until the condition of resonance is reached. A spectrum is subsequently produced relating the amount of energy absorbed from the oscillating field to the chemical shift δ.

The essential components of an NMR spectrometer (Fig. 17.1) are:

1. A strong magnetic field that can be varied continuously and precisely over a relatively narrow range;
2. A radiofrequency transmitter and a radiofrequency receiver;
3. A sample holder that positions the sample relative to the main magnetic field, the transmitter coil and the receiver coil;
4. A V/F converter and a computer for recording the data.

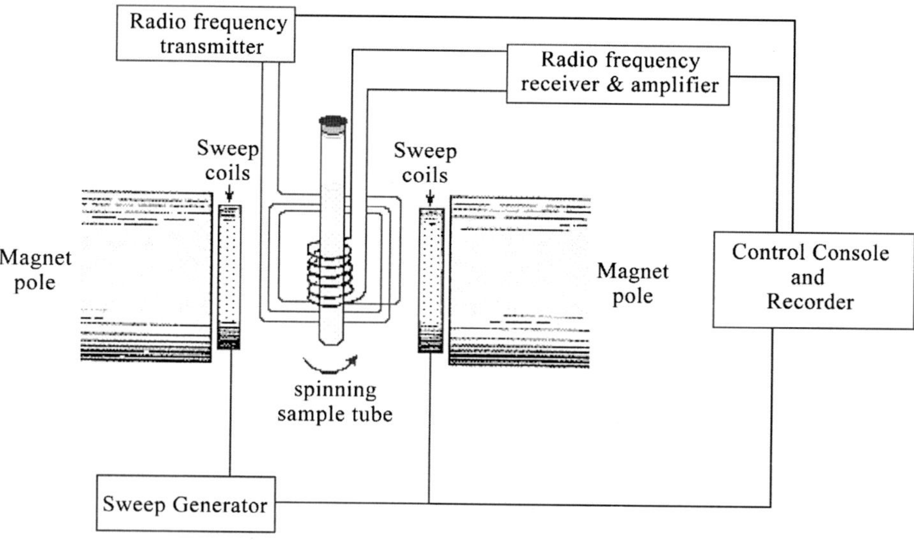

Fig. 17.1: Block diagram of a typical NMR spectrometer

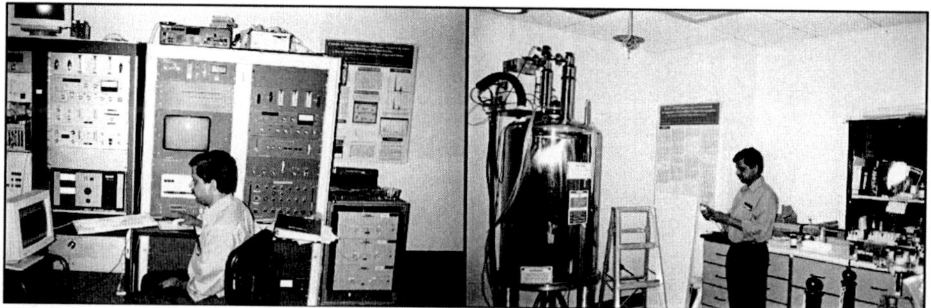

Fig. 17.2: ^{13}C-NMR spectrometer (Author at University of Florida)

The sample is placed in the sample holder of the spectrometer and the magnetic field H_0 is increased, thereby causing the precession rate of the nuclei to increase. When the frequency of the alternating magnetic field H_1 is matched by the precession frequency, the nuclei flip over and a voltage is induced in the detector coil. This resonance signal is subsequently amplified and recorded. From the spectrum that is produced, information is provided regarding the chemical environment of C or H atoms in the sample.

Chemical shift δ is the parameter from which structural information is obtained. The basis for the analysis is that the frequency at which nuclei resonate is governed by the chemical environment of the nuclei in the sample. For example, the chemical shift δ of ^{13}C in an aromatic ring is different from that of a COOH group; thus, the two can be separated by NMR spectroscopy.

17.2. Techniques for ^{13}C-NMR Spectroscopy

The technique of ^{13}C-NMR spectroscopy (Fig. 17.2 and 17.3) is currently the most widely used and definitive instrumental technique for the characterization of organic matter. Two general approaches have been used: liquid state ^{13}C-NMR spectroscopy and solid state ^{13}C- NMR spectroscopy. Because of the insolubility of some components of soil organic matter (humin), greater use has been made of solid-state ^{13}C-NMR spectroscopy. Also, solid state ^{13}C-NMR spectroscopy is suitable for the examination of organic matter *in situ*.

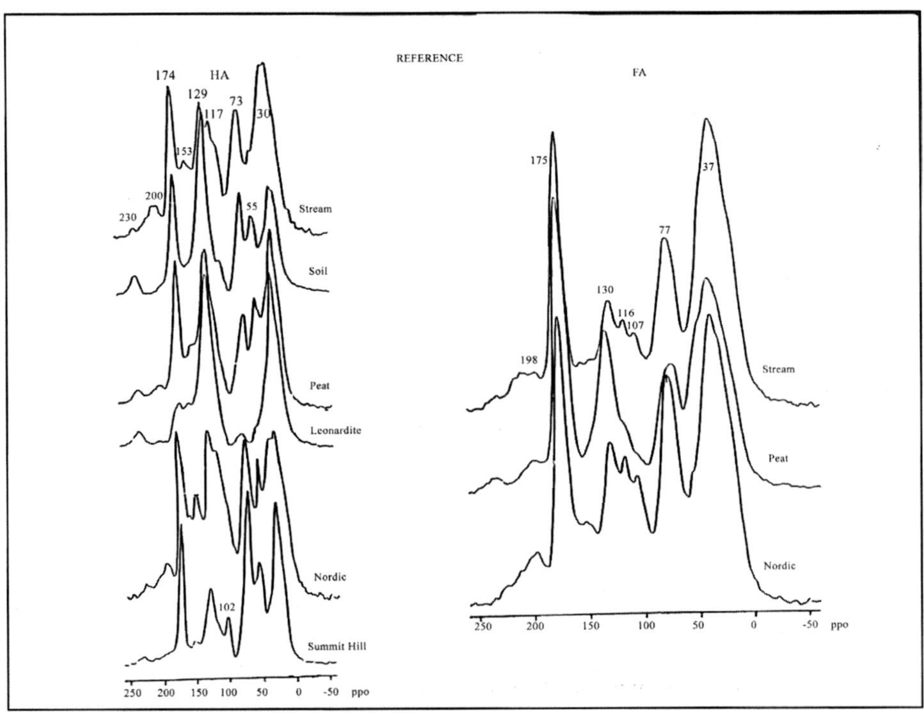

Fig. 17.3: ^{13}C-NMR spectra for humic acid and fulvic acid reference standards

17.3. Limitations of ^{13}C-NMR Spectroscopy

Some limitations of ^{13}C-NMR spectroscopy are listed below:

1. The natural abundance of ^{13}C is very low (~ 1.1% of total organic carbon). The most abundant isotope of carbon (^{12}C) does not exhibit a magnetic spin momentum and thus cannot be observed by NMR.
2. It is the inherent nature of the ^{13}C-NMR experiment that a very large number of signals must be collected and averaged before an adequate response can be measured and recorded. Accordingly, long analysis times are required.

3. For solids, line broadening and other effects reduce resolution.
4. Costs for the puchase and maintenance of a modern ^{13}C-NMR spectrometer may be beyond the means of most laboratories.
5. As applied to humic substsnces, results of ^{13}C-NMR spectroscopy must be interpreted with caution, because "chemical shift" regions are not completely exclusive or specific.

17.4. Advantages and Disadvantages of Liquid State ^{13}C-NMR Spectroscopy

Most liquid state ^{13}C-NMR spectra of humic substances have been obtained in an alkaline solution (0.1 *M* NaOH) with sample concentrations of 50 to 100 mg mL^{-1} (100 to 200 mg in sample analyzed) on NMR spectrometers with resonance frequencies ranging from 200 to 400 MHz. The spectrometer time required to obtain a spectrum of satisfatory signal-to-noise ratio depends on the type of experiment but will normally be of the order of 6 to 12 hours.

Liquid-state NMR has both advantages and disadvantages. One major advantage is decreased line broadening due to dipolar interactions and chemical shift anisotropy. Some disadvantages are:

1. Low solubility of humic substances in suitable NMR solvents. The differential solubility of various components in a mixture can result in sample fractionation.
2. Formation of micelles and colloidal suspensions.
3. Long relaxation times and proton coupling effects.
4. Solvent effect on chemical shifts. Chemical shifts of the sample can also be obscured by chemical shifts due to the solvent.
5. Inability to cool the sample to low temperatures and still maintain a liquid phase.

17.5. Limitations of Solid-State ^{13}C-NMR Spectroscopy

An undesirable feature of solid-state ^{13}C-NMR is that of line-broadening due to:

1. Dipolar interactions between ^{13}C and ^{1}H nuclei;
2. Chemical shift anisotropy of highly anisotropic carbons (primary aromatic and carboxyl);
3. A poor signal-to-noise ratio.

These problems are not encountered in liquid-state ^{13}C-NMR. One advantage of soild-state is that the sample can be recovered in an unaltered state.

The utility of solid-state ^{13}C-NMR has been greatly enhanced in recent years by using a combination of the following techniques to improve spectral quality:

High-power proton decoupling: Humic substances contain two interacting spin species - ^{13}C and ^{1}H. Much of the ^{13}C line broadening arises from interactions between the two. Through high-power proton decoupling, the magnetic influence of the proton nuclei on a neighbouring ^{13}C nuclei is eliminated or minimized, thereby reducing line broadening.

Cross-polarization: This technique results in the transfer of net magnetization from the abundant ^{1}H spins to the less abundant ^{13}C spins. The procedure also overcomes the problem of dipolar line broadening; in addition, resolution is enhanced by an increase in net ^{13}C magnetization.

Magnetic-angle spinning: This is a technique for further decreasing the broadening by eliminating the remaining vestiges of dipolar ^{13}C–^{1}H interactions and chemical shift anisotropy effects by rapidly rotating the sample at the so-called "magnetic angle" with respect to the applied magnetic field. The magnetic angle is 54.7°.

All the three techniques must be used simultaneously in order to obtain optimum results.

17.6. Typical Applications of ^{13}C-NMR Spectroscopy

The following are some of the typical applications of ^{13}C-NMR Spectroscopy:

1. Studying the chemical composition of humic acid, fulvic acid and humin fractions of soil organic matter, mineral soils and their size and density fractions, histosols, forest litter and humic materials in sewage sludge;
2. Determining the cultivation effects on soil organic matter;
3. Characterizing the transformations of organic matter;
4. Examining the pathways of humus synthesis;
5. Monitoring the chemical changes in composted organic matter;
6. Determining the contribution of crop residues to total soil organic carbon.

18

Nuclear Techniques for Soil Organic Matter Studies

The study of soil organic matter is becoming increasingly important by the day as world agriculture attempts to ensure sustainability of soils, while at the same time striving to enhance food production to feed an ever-burgeoning population. The use of green manures and pasture leys, the return of residues and additions of organic amendments to the soil are often resorted to in an attempt to increase soil organic matter which immensely benefit the physical, chemical and biological make-up of the soil.

There has been an ever-growing interest among researchers on the possibilities of using isotope and nuclear techniques for soil organic matter studies. Application of techniques such as radiocarbon dating and stable isotope ratios have opened up whole new vistas of research. The key areas where isotope techniques have found unique role include the effects of organic matter on nutrient availability, the synthesis and decomposition of organic matter in soils, the chemistry of humic substances and the transformations of nitrogen, phosphorus and sulphur in soils.

It is not within the ambit of this book to present a complete review of the studies on soil organic matter made with the use of nuclear techniques. What is attempted here is to highlight the potential areas of soil organic matter research possible with nuclear techniques in order to stimulate students and researchers who have not yet considered the use of radioisotopes and nuclear techniques in their research.

18.1. Preparation of ^{14}C and ^{3}H Labelled Plant Materials

Preparation of ^{14}C labelled experimental material is essential for the studies on the effect of microorganisms on the decomposition of soil organic matter and of the humification of plant and microbial tissues. The normal procedure to produce uniformly ^{14}C labelled plants is to grow them in a closed system enriched with $^{14}CO_2$. Several successful approaches have been adopted, based on sealed plant growth chambers, ranging in complexity, from the very elaborate $^{14}CO_2$ growth chambers[2] to simple polystyrene canopies[3]. Essential features include complete sealing from the outside atmosphere, provision of continuous air circulation for producing, maintaining and monitoring the desired level of $^{14}CO_2$ and control of ambient temperature. Watering and supply of nutrient solutions must also be provided.

Normally, closed systems will have 0.1% $^{14}CO_2$ by volume in the system, produced by reacting required amount of labelled $Na_2{}^{14}CO_3$, with dilute mineral acid to produce an activity of around 100 µCi of ^{14}C per gram of ^{12}C. The final specific activity of labelled plant carbon will be about one-third of that of $Na_2{}^{14}CO_3$.

The possibility of radiation damage to the plants imposes a limit on the activity of ^{14}C that can be used in the system, but the objective is to obtain labelled plant material containing 25 - 100 µCi of ^{14}C per gram of ^{12}C. Plant materials with this amount of activity can be used for studies on the chemistry of humic substances, although less activity may be enough for studies where specific fractionation or separation is not contemplated. Normally, fast growing species such as wheat or barley is used. It is imperative that the materials should be uniformly labelled which entails that lignin and cellulose should have the same specific activity as the water-soluble fractions.

Several endeavours have been made to simplify methods by providing the ^{14}C label to the roots by means of labelled cyanamide fertilizer $H_2{}^{14}CN_2$ or as labelled $NaH^{14}CO_3$. However, these methods raise problems of their own and a sealed growth chamber may still be required to prevent the loss of $^{14}CO_2$ through dark respiration, resulting in radioactive contamination and loss of valuable label. Also important is very large CO_2 dilution that would occur in plants exposed to normal atmosphere.

Normally, ^{14}C is introduced into soil organic matter and humic substances by providing labelled plant material as a substrate for organisms already present in the soil. Neutron irradiation has been used to label humic acids, but is not practical due to lack of specificity, the unknown nature of the products and the presence of radiation decomposition products.

Tritiated ^{3}H humic substance can be prepared by allowing plant material to decompose in soil to which ^{3}H-labelled water has been added.

---

[2]Zeller, A., Oberlander, H. E. and Hiemer, F. A. 1963. A growth chamber for raising ^{14}C labelled plants. *In: Use of Isotopes in Soil Organic Matter Studies*. FAO/IAEA Tech. Meeting. Pergamon, Oxford. p. 401.

[3]Warembourg, F. R. and Paul, E. A. 1973. The use of $^{14}CO_2$ canopy technique for measuring carbon transfer through the plant-soil system. *Plant and Soil* 38: 331-345.

18.2. Assimilate Transfer and Organic Matter Production

By exposing plants to $^{14}CO_2$ under a cellulose acetate-butyrate canopy, non-uniform labelling of growing plants in the field has been carried out. The objective here is to determine the carbon transfer through the plant-soil system by measuring the carbon assimilation by plants, the transfer of photosynthate below ground and the respiration of the roots.

Studies on root formation and decomposition during plant growth, with plants grown for several months in a $^{14}CO_2$ atmosphere (with shoots sealed from roots), have proved that plant roots are of much greater significance as a continual source of both soil humus and microbial energy than are generally recognized. It was also known that residual quantity of roots at harvest cannot give a complete view of the role of roots in organic matter turnover and formation.

Plants grown in a $^{14}CO_2$ atmosphere have also indicated that a major part of the organic ^{14}C present in the soil resulted from autolysis of root tissue. Such $^{14}CO_2$ studies will do much to improve our knowledge of root metabolism, breakdown and the formation and maintenance of organic matter.

18.3. Influence of Soil Organic Matter on Soil Fertility and Chemical Residues

Organic matter, besides being itself an important source of plant nutrients, wields considerable influence on the availability of plant nutrients from the soil and on plant nutrition and metabolism. It has also an effect on the exchange capacity of soils, the uptake of fallout nuclides, such as ^{137}Cs, and on the inactivation of herbicide and pesticide residues. The range of potential studies is very vast, covering the whole gamut of the use of isotopes in soil research.

Fried and Dean (1952)[4] pointed out that when two sources of nutrients are present in the soil, the plant would absorb from each of these sources in proportion to the respective quantities "available". Thus, the amount of available nutrient in the soil can be determined in terms of a standard, provided the proportion of the nutrient in the plant derived from this standard is determined:

$$A = \frac{B\,(1-y)}{y}$$

[4]Fried, M. and Dean, L. A. 1952. A concept concerning the measurement of available soil nutrients. *Soil Sci.* 73: 263-271.

where,

A = amount of available nutrient in the soil

B = amount of fertilizer nutrient (standard) applied

y = proportion of nutrient in the plant derived from the standard

Although 'A' value was primarily developed to determine the availability of phosphorus in soil, the principles employed are now applied to a number of nutrient elements including nitrogen (using ^{15}N), sulphur (^{35}S), calcium (^{45}Ca) and zinc (^{65}Zn).

It will be appreciated that the use of 'A' value is not confined to estimating availabilities of nutrients in soil, but can also be applied to evaluate the relative availability of different phosphorus sources on the same soil and to determine the residual value of applied phosphorus. For instance, the availability of phosphate residues in three calcareous soils has been evaluated from 'A' value measured by subtracting the 'A' value of untreated plot from that for the treated. Also, the effect of long-time dry land cropping on available phosphorus in a fine sandy loam has been studied.

While 'A' value is most often used to measure amounts of available nutrients in different soils or fertilizer materials, it can also be used to measure indirectly the availability of nutrients from sources that cannot be labelled conveniently. Such sources include rock phosphate and residues from organic materials, or from previous application of fertilizers and organic manures[5].

The influence of soil organic matter on available phosphate in soil has been studied with the radionuclide ^{32}P. The problem of the rapid removal of zinc from the available pool when zinc is added to 'zinc deficient', high organic matter soils has been studied using ^{65}Zn. It was feasible to pursue changes in the zinc distribution as reflected by different extractants and in the organic matter residues. Adding organic matter to the soil resulted in more zinc going into the organic fraction and zinc deficiency was caused during the active decomposition of organic matter. Studies with ^{54}Mn and ^{59}Fe have shown that presence of organic matter is one of the important factors responsible for retaining high concentrations of soil solution manganese and that organic matter helps in the mobility of iron and manganese in calcareous soils. The chelating effect of organic matter has also been found to have an essential role in the migration of fission products.

---

[5]Sangeeta Mohanty, Narendra Kumar Paikaray and A. Raja Rajan. 2006. Availability and uptake of phosphorus from organic manures in groundnut (*Arachis hypogea* L.) – corn (*Zea mays* L.) sequence using radio tracer technique. *Geoderma* 133: 225-230.

18.4. Role of Microorganisms in Decomposition of Soil Organic Matter

With ^{14}C compounds, it has been easier to determine factors influencing the turnover of organic matter in soil, the effect of adding fresh organic matter (residues) to the decomposition of native carbon soils and the general rapid estimation of microbial activity.

Often the use of ^{14}C labelled material is an essential way of tracing the decomposition of added organic matter. In priming studies, ^{14}C labelled organic matter is incorporated into the soil and by comparing the specific activity of the evolved $^{14}CO_2$ with that of the added organic matter, it is possible to calculate what proportion of the mineralized carbon comes from the added organic matter and what comes from the decomposition of the native organic matter. Such typical isotope dilution procedures have largely contributed to the present views that priming effect is largely a turnover phenomenon. General microbial activity can be determined by a relatively simple process of measuring $^{14}CO_2$ evolution and this can be a measure of the potential productivity of a given soil.

There are a number of references concerning the use of isotopes in studies on the decomposition of organic matter. The use of specifically ^{14}C labelled lignins, coniferyl alcohols and humic acid-type polymers has given information about their breakdown in soils. Tracer techniques with ^{14}C and ^{15}N have made it possible to determine readily the turnover time of the less resistant soil constituents, showing that there is an early rapid breakdown of plant residues when placed in the soil, but there is a resistant fraction of principally microbial metabolites and cell wall constituents which stabilizes and remains over a period of years.

Determining the turnover rate or mean residence time of the stable fraction has been made practicable by using the naturally occurring ^{14}C for radiocarbon dating. This technique can be combined with tracer techniques to give a complete picture of the mean residence time of different soil organic matter constituents[6].

18.4.1. Determination of Soil Microbial Activity by Radiorespirometry

This rapid procedure provides a measure of the microbial activity of soils. It can be used to compare soils as an estimate of their inherent fertility, or can be used to evaluate the possibly transient effects of the application of a pesticide or of some soil treatment.

The principle of this method is that if a ^{14}C labelled substrate is presented to the sample and the transformation rate is determined by $^{14}CO_2$ evolution, then this will be a direct measure of microbial population and activity.

Simple respirometer flasks are prepared, consisting of 100 mL conical flasks fitted with rubber stopper into which is inserted a glass rod carrying a small polythene or glass vial. The vials can be attached to the rods with an adhesive. In the vial a strip of pleated filter paper, impregnated with 0.25 mL of phenethylamine as a CO_2 absorbent, is placed. A soil sample (2 or 3 gram) is placed in each flask, moistened with enough water containing 0.5 mL of a ^{14}C glucose solution to keep it at optimum moisture content. The activity of the ^{14}C glucose solution should be in the range of 0.01 – 0.04 μCi (370 – 1480 dps). The stopper is inserted and the sample is incubated at 25 °C for a desired time at the end of which the reaction is terminated by adding 2 - 3 mL of 1 *N* HCl. The filter paper and the remaining phenethylamine are then transferred quantitatively to a scintillation vial, 10 mL of pertinent scintillation cocktail added and shaken well prior to counting. By comparison with a standard, the μmoles of glucose consumed per hour per gram of oven dry soil is calculated.

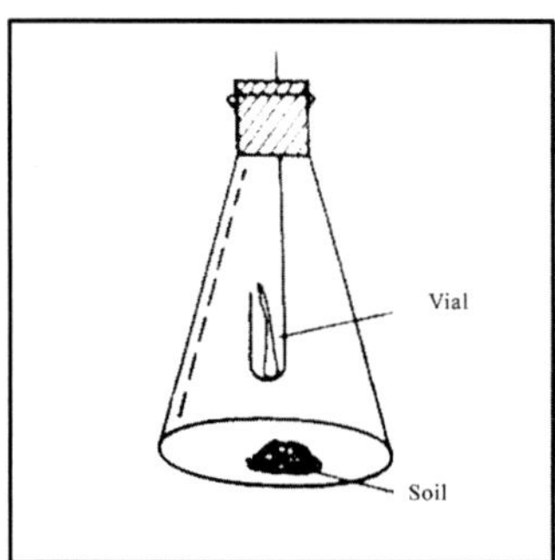

Normally, the samples are replicated, with incubation times of 30, 60 and 90 minutes and with different substrate concentrations, such as 0.5, 1.0, 5.0 and 10 μmol glucose per 0.5 mL solution. This method is reproducible and provides an accurate and useful measure of soil microbial activity.

18.4.2. Determination of Turnover Rate of Crop Residues

It might be interesting to determine the fractional rate at which organic matter pool decomposes and is renewed. Under steady state conditions, the amount of soil organic matter lost over a period of time will equal the amount added. The fractional rate of change is constant which means it is exponential. Therefore, a semi-log plot of log carbon added through residues versus time on a linear scale will be a straight line (Fig. 18.1).

[6]Paul, E. A. and McGill, W. B. 1977. Turnover of microbial biomass, plant residues and soil humic constituents under field conditions. *In: Soil Organic Matter Studies*. Proc. Symp. Braunschweig, 1976. IAEA, Vienna.

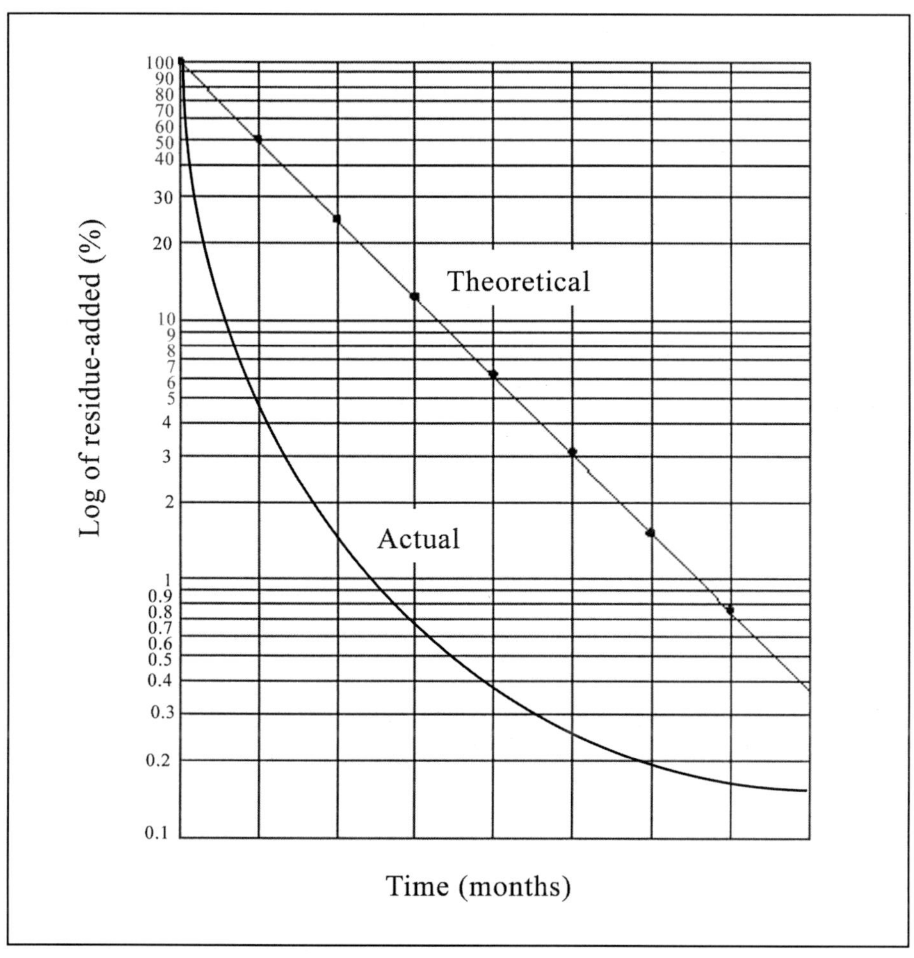

Fig. 18.1: Decomposition of added plant residues with time

It can be assumed that:

$$2.3 \log \frac{Q}{Q_0} = -kt \longrightarrow \text{Equation (1)}$$

$$\text{Or, } Q = Q_0 e^{-kt} \longrightarrow \text{Equation (2)}$$

where

Q_0 = Quantity of organic matter present at zero time

Q = Quantity of organic matter present at time t

e = base of natural logarithm = 2.7183

Then, the equation (1) above, solved for biological half-life $t_{1/2}$ is:

$$t_{1/2} = \frac{0.693}{k}$$

Therefore, the turnover rate constant, $k = \frac{0.693}{t_{1/2}}$

The principle of the method is that ^{14}C labelled plant material is added to microplots (about 1 m^2) in the field. Then at successive periods of time, every few weeks or months, the plot is sub-sampled and the amount of ^{14}C from the labelled material still residual in the soil is measured as $^{14}CO_2$. If desired, the ^{14}C of any particular soil organic fraction can be estimated after fractionation by one of the existing procedures (*vide* Chapter 20).

18.5. Radiocarbon Dating of Soil Organic Matter

The objective of natural radiocarbon dating of soils is to evaluate the long-term organic matter turnover of soils and thus indirectly provide a means of estimating the nitrogen supplying power of soils.

Radiocarbon dating procedures are based on the fact that high energy neutrons from the cosmic radiation react constantly with the nitrogen of earth's atmosphere to yield ^{14}C by this nuclear reaction: ^{14}N (n, p) ^{14}C. The resulting ^{14}C combines with atmospheric oxygen to form radioactive carbon dioxide, which is incorporated into plants by photosynthesis; animals then acquire ^{14}C by eating the plants. Once a living organism dies, taking in new carbon is stopped. The ratio of ^{12}C to ^{14}C at the moment of death is the same as every other living thing, but the ^{14}C decays and is not replaced. The ^{14}C decays with a half-life of 5,730 years, while the amount of ^{12}C remains constant in the sample. By looking at the ratio of ^{12}C to ^{14}C in the sample and comparing it to the ratio in a living organism, it is possible to determine the age of a formerly living thing fairly precisely. The average specific activity of modern carbon is 13.56 ± 0.07 dpm g^{-1} of C. There is about 75 t of radiocarbon on earth. The isotope ratio of ^{14}C to ^{12}C is 1.2×10^{-12}.

There are two assumptions in radiocarbon dating: when a living organism dies, the carbon atoms of the dead matter do not exchange with external carbon such as CO_2 of the atmosphere; secondly, the specific activity of the ^{14}C in the biosphere has reached a steady state. It then follows that if a plant material died t years ago ($t = 0$), it had then a specific activity of ^{14}C equal to S_0. If organic matter from this material is isolated today (t years later) and the present specific activity S is determined, then t can be calculated from the decay equation:

$$S = S_0 (½)^{t/t½}$$

where $t_{½}$ is the half-life of ^{14}C, S_0 is the specific activity of a standard reference and t is the mean residence time of the organic C in the soil.

The very low levels of naturally occurring ^{14}C (a weak beta emitter) entail significant problems of technique to get a representative sample, especially to have sufficient activity for measurement. As the activity is low, the CO_2 evolved is converted into acetylene (C_2H_2) for gas proportional counting or the acetylene is converted into benzene to which is added a toluene-based scintillation mixture for liquid scintillation counting.

Though procedures may vary, in general, a soil sample containing 3 or more g of organic carbon is taken and sieved to remove roots and recognizable plant material pieces, dried and ground to pass through a 60 or 80 mesh sieve. The sample must be kept acidic (pH 6.0) to avoid adsorption of atmospheric CO_2. The sample is shaken in 0.1 *N* HCl (1:15 ratio) and centrifuged at 2000 rpm for 15 minutes and the supernatant discarded. This is repeated, if necessary, to remove undecomposed organic matter, followed by drying. Inorganic carbonate is removed by boiling in 1 *N* HCl for 5 minutes and the sample is dried by evaporation. Combustion in a furnace yields CO_2 which is absorbed in 200 - 250 mL of 25% NH_4OH (or, NaOH, precipitating $CaCO_3$ by addition of saturated $CaCl_2$). In case of the former, the solution can be poured into 500 mL of 20% $SrCl_2$ solution to form $SrCO_3$ which is cooled, filtered, washed and dried in vacuum to prevent atmospheric CO_2 contamination.

The carbonate can then be taken for conversion into acetylene for proportional gas counting or further conversion to benzene, or to CO_2 for ratio analysis of $^{12}C/^{13}C$ by mass spectrometer.

Typical data show mean residence times of some organic fractions from 300 to over 5,000 years, the latter being from 95 cm depth. Residence times of 1,000 to 1,500 years have been reported common for non-hydrolysable humic fractions, as compared to 10 - 15 years for hydrolysable humic fractions.

18.6. ^{13}C and ^{14}C Isotope Studies

Carbon is the energy source, driving many of the nutrient cycles that occur in the soil. A ready supply of accessible carbon is necessary for a continuous supply of soil nutrients and to maintain soil structure.

It can often be difficult to study the effect of these procedures on the SOM pools and soil chemical and physical fertility because of the large amount of background carbon present in the soils. The use of carbon isotopes offers an easy and precise method of tracing the additions of different plant materials to soil carbon fractions and their influence on such soil properties as soil nutrients and structure.

Bingeman *et al.* (1953)[7] and Hallam and Batholomew (1953)[8] were some of the first researchers to use ^{14}C in soil chemistry studies of soil organic matter. Since then, ^{14}C has been used in numerous aspects of soil organic matter research. By using substrates that were labelled with ^{14}C and/or ^{13}C it became possible to trace the decomposition of added plant residues with considerable accuracy, even when there were relatively large amounts of native organic matter present. By using ^{14}C labelled materials it was possible to identify plant C as it became incorporated into different fractions of the soil humus. ^{14}C labeled corn residues have been used to study the carbon dynamics and respiration of soil organisms in soils under different tillage regimes and to estimate the carbon budget and carbon partitioning during a short period after residue application.

Plant roots grown in a ^{13}C-depleted CO_2 atmosphere were used to study the amount of C mineralized from decomposing wheat roots grown under ambient and elevated CO_2 concentrations. The plant roots had $\delta^{13}C$ values of -38.3‰ and -41.1‰. By using these ^{13}C-depleted roots they were able to distinguish between the root-derived C and native soil C that was mineralized. The use of ^{13}C labelled wheat straw made it possible to determine the fate of organic matter in water stable soil aggregates of differing sizes during the decomposition of newly added crop residues. Wheat plants were grown in a growth chamber containing 2% ^{13}C-CO_2 to produce a uniformly labelled straw with an enrichment of 9.060 atom% ^{13}C excess. Oilseed rape labelled with ^{13}C have been used to investigate the decomposition of the residues added to soil and to determine how the breakdown of the plant materials related to N content, soluble carbon compounds and the cellulose and lignin content of the plant material. The labelled plant material was obtained by transferring 2-4 leaf stage spring oilseed rape plants to an air-tight growth chamber. The plants were then grown with continual labelling of $^{13}CO_2$ with an isotopic excess of 3.13%, throughout the whole growth cycle.

---

[7]Bingeman, C. W., Varner, J. E. and Martin, W. P. 1953. The effect of the addition of organic materials on the decomposition of an organic soil. *Soil Sci. Soc. Am. Proc.* 17: 34-38.

[8]Hallam, M. J. and Bartholomew, W. V. 1953. Influence of rate of plant residue addition in accelerating the decomposition of soil organic matter. *Soil Sci. Soc. Am. Proc.* 17: 365-368.

18.6.1. Natural Abundance of Carbon Isotopes

The ratio of ^{13}C to ^{12}C in the atmosphere can vary with different physiographic parameters such as altitude, latitude and temperature as well as by some biological processes. When plants fix carbon during photosynthesis there is a subtle degree of discrimination between the amount of ^{13}C and ^{12}C. Discrimination occurs during the carboxylation step in photosynthesis, with the discrimination against ^{13}C being greater in C_3 (Calvin cycle) plants than in C_4 (Hatch-Slack cycle) plants, due to the greater discrimination in the primary carboxylation step of C_3 plants. This primary carboxylation step is catalyzed by the enzyme ribulase biphosphate carboxylase (RuBP) resulting in a lower ^{13}C:^{12}C ratio in C_3 plants than in C_4. CAM plants (Crassulacean acid metabolism plants) show variable discrimination, but it is more often similar to C_4 plants.

The ^{13}C:^{12}C ratio is generally measured as $\delta^{13}C$. A C_4 species such as maize will have a $\delta^{13}C$ value of approximately –12 per mil (‰) whereas, in a C_3 species such as wheat or rice it will be approximately –26‰. The $\delta^{13}C$ of soil organic matter is comparable to that of the source plant material and thus every change in vegetation between C_3 and C_4 plants results in a corresponding change in the $\delta^{13}C$ value of the soil organic matter. This means that when C_3 plants are grown in soils, which had previously been under C_4 vegetation (or *vice versa*), there is virtually an *in-situ* labelling of the organic matter incorporated into the soil.

18.6.2. Measurement of ^{13}C

The ^{13}C is most often determined in CO_2 produced from a solid sample combusted in a stream of oxygen. The equipment most commonly used are the Leco and Carlo-Erba furnaces linked to a mass spectrometer set to measure the mass 45/44 ratio. The results are expressed as $\delta^{13}C$ (‰), which is not the absolute isotope ratio but that relative to a standard. The original standard used was a limestone fossil of *Belemnitella americana* (PDB) from the cretaceous Peedee formation in South Carolina, USA. Because this material is no longer available, other cross calibrated standards are used.

18.6.3. Added Residues Remaining in the Soil

The proportion of soil C derived from the C_3 (or C_4) plant can be calculated from the equation:

$$\chi = \frac{\delta f / \delta s}{\delta r / \delta s}$$

where δf is the $\delta^{13}C$ value of the soil at time t after the addition of the residues, δs is the $\delta^{13}C$ of the original soil or soil of the control treatment and, δr is the $\delta^{13}C$ of the C_3 (or C_4) plant residue added to the soil.

If the total C content C of the soil is known then the absolute quantity X of carbon from the C_3 (or C_4) plants can be determined from the equation:

$$X = \chi \times C$$

The absolute quantity Y of residual carbon from the initial soil can be determined from the equation:

$$Y = C (1 - \chi)$$

18.7. Laser CO_2 Carbon Isotope Analyzers

For enhancing food security and sustainable agriculture, agricultural management practices that optimize soil organic carbon sequestration and water and nutrient use efficiency are important. These practices also promote climate-smart agriculture by making soils and plants more resilient to climate changes and by reducing greenhouse gas emissions.

Different agricultural practices have different impact on CO_2 release or capture affecting CO_2 concentrations in the air. Ploughing, for example, can stimulate the decomposition of organic matter in soil, leading to the release of more CO_2, while the use of mulch could help soil and microbes to store CO_2. Accurate measurements and analysis of data are the first requirements in ensuring proper evaluation and selection of such agricultural management practices.

In this endeavour, laser CO_2 carbon isotope analyzers are used to evaluate agricultural management practices and optimize soil organic carbon sequestration for sustainable and climate-smart agriculture. Compared to conventional methods, such as gas chromatography and isotope ratio mass spectroscopy, the laser CO_2 carbon isotope analysis is a relatively new technology that allows for real-time, *in situ* measurements of the concentration and ^{13}C signature of CO_2 emitted from soil, microorganisms and plants. Because this type of analysis is relatively simple and can provide robust data, it is increasingly being used to monitor and track CO_2 emissions.

For more information, readers can refer to International Atomic Energy Agency's Tecdoc Series no. 1866 which gives a treatise on the specific protocols on the measurements, instrument calibration and data management in the use of laser CO_2 carbon isotope analyzers.

19

Managing Soil Organic Matter

Managing soil organic matter is of paramount importance in sustainable crop production. Human interventions always lead to a decrease in soil organic matter content and biological activity. Therefore, even for maintaining moderate levels of organic matter in soils, sustained efforts should be in place. Such sustained efforts should include adopting high-residue crops in rotation and ensuring regular return of residues to soils.

It is all the more difficult to maintain the organic matter content in well aerated soils and soils of hot and arid regions, because in these soils added materials decompose rapidly. However, in comparison, it may not be so difficult to maintain the organic matter levels in fine-textured soils with restricted aeration and in soils of cold temperate regions.

19.1. Measures that Decrease Soil Organic Matter Content

Human interference of any kind and magnitude disturbs the equilibrium of soil ecosystem through its influences on soil organisms. Measures such as burning vegetation or repetitive tillage can alter the living status of soil organisms and degrade their ecology. Eventually, this can lead to a reduction in soil biomass and biodiversity.

When the soil is devoid of organisms that decompose soil organic matter and bind soil particles, soil structure is destroyed by rain and wind. Destruction of soil structure can lead to rainwater runoff and soil erosion, removing the organic matter of the topsoil which is the potential food for the organisms. Therefore, the most important component of soil ecology is soil biota, bereft of which a soil ceases to be soil.

A reduction in soil organic matter content is brought about by a set of factors that result in:

- decreased biomass yield;
- decreased organic matter stock;
- increased decomposition toll of soil organic matter.

19.1.1. Decreased Biomass Yield

a. Deforestation

Deforestation of a land for agricultural purposes leads to the loss of the litter layer which substantially reduces the population and diversity of soil organisms. Many temperate forest species can adapt well to grassland situations and, therefore, the effects of deforestation are far more pronounced in the tropics than in the temperate regions. When soil biodiversity declines and adapted species supersede the indigenous species, drastic changes in the composition of the macrofaunal community take place. When indigenous species disappear, only adapted species are then left behind for recolonization.

b. Monoculture

Replacement of mixed vegetation (Fig. 19.1) with monoculture results in a decrease in faunal diversity under situations of simplified vegetation and the disappearance of the litter layer under grassland and single crop systems. Even though root systems, especially of grasses, can extensively forage vast areas of soil, root exudates from a single crop will never be able to attract a multitude of divergent species of microorganisms. Thus, predator diversity may be affected. Relatively more opportunistic pathogen species will gain proximity to the crop and harm the crop.

Fig. 19.1: Replacement of mixed vegetation with monoculture leads to a decrease in faunal diversity

Long-term cultivation and grazing will lead to compacted layers of soil, affecting soil porosity and circulation of soil air. Anaerobic conditions in the soil will invigorate the development of more pathogenic organisms.

c. Bare Fallows

In conventional agriculture, a fallow period is advocated following a period of crop production to "rest" the land and to rejuvenate its productivity. Usually, this is imperative in conventional tillage systems characterized by depleted nutrient supply and altered soil biota. When a land is left as bare fallow (Fig.19.2), the soil biota is left with no energy source, except for the spontaneous, sparse weed population. Thus, the soil food web is not being recovered and, additionally, soil organic matter is also degraded. Also, the lack of crop cover could accentuate runoff and erosion during the rains.

Fig. 19.2: A bare fallow land could result in loss of SOM and severe run off and erosion

19.1.2. Decreased Organic Matter Stock

a. Burning of Crop Residues

Burning of crop residues in the field is a common practice. This is usually intended to control pests and diseases and to render fieldwork easier in the following season. Burning results in the destruction of the litter layer and in the diminishing of the quantity of organic matter that is returned to the soil. In the process, the habitant organisms on the surface soil are destroyed. For the future decomposition of organic matter and release of plant nutrients to take place, energy has to be expended first in the restoration of the microbial community.

b. Excessive Grazing

Throughout the world, grazing lands are tended to be overstocked, beyond their bearing potential. Overgrazing by draught animals, cows and small ruminants on community lands, river banks and other public utility places could destroy the most palatable species in the plant population and reduce the density of plant cover. This could enhance the erosion hazard and reduce the nutritive value and bearing potential of the land.

c. Unreplenished Crop Residues

Farmers remove crop residues from the field to be used as fodder or bedding material for cattle or for composting. If these residues are subsequently returned to the soil as manures or composts, they enrich soil fertility. Sometimes, crop residues removed from the field are not returned. This results in the impoverishment of the soil in terms of soil fertility and recycling of plant nutrients that constitute the residues becomes untenable.

19.1.3. Increased Decomposition Toll of Soil Organic Matter

a. Tillage Operations

When a soil is tilled, the soil is opened up and aerated. The residues are incorporated in the soil along with air. Oxygen invigorates the soil microorganisms that feed on organic matter and the decomposition of organic matter is fast paced, leading to the formation of stable humus and release of CO_2 to the atmosphere. Thus, tillage results in a reduction in soil organic matter. The more a soil is tilled, the more is the organic matter that is lost from the soil by way of decomposition.

In contrast, the residues lying on the soil surface (without incorporation in the soil) are exposed to fewer microorganisms and therefore decompose relatively slowly and release less CO_2 into the atmosphere.

Conventional tillage not only decreases organic matter conservation in soil by accelerating its decomposition, but it also does increase the potential loss of valuable soil by water and wind erosions. The impacts could be listed thus:

- When a soil tilled, residues are not left on the soil surface to overwhelm the impact of rains.
- Tilling reduces the food resources available for earthworms leading to a reduction in their population and activity; this, aside from reducing soil aeration and porosity, impacts the incorporation of plant residues in soil, favouring rapid decomposition of organic matter.

- Repetitive tillage destroys soil aggregates and disconnects the innumerable channels connecting the surface soil with the subsurface soil; once the large pores are destroyed by conventional tillage practices, rain water is not ushered into the soil.
- A compacted layer of plough pan develops due to the action of tillage and this hampers root penetration and water infiltration.
- Ploughing a dry soil worsens the pulverization of soil, leading to greater water runoff and erosion; greater water runoff portends stress due to drought, later in the season.

b. Drainage

Soil drainage is primarily influenced by the topography of the land. Water tends to stagnate in soils in low-lying areas. Also, frequent use of tillage implements develops subsoil impervious layers that impede free drainage, resulting in water-logged conditions.

In wet soil conditions, there is bound to be an accumulation of organic matter because organic matter decomposes more slowly in poorly-aerated soils than in well-aerated soils. Lignin does not decompose at all in a permanently waterlogged soil. In extremely swamp environments, organic soils (peat or muck) develop, with organic matter content exceeding 30%. If wet soils are drained artificially, the decomposition of organic matter can then take place rapidly. A well-drained soil is well aerated and favours decomposition of soil organic matter.

c. Agrochemicals Use

Generally, agrochemicals such as fertilizers and pesticides enhance crop growth and biomass production. However, application of nitrogenous fertilizers and pesticides can trigger the activity of microorganisms and, thus, the decomposition of organic matter. This is especially significant in situations where the C:N ratio of the soil organic matter is high and the decomposition rate of organic matter is poor due to lack of nitrogen.

19.2. Measures that Increase Soil Organic Matter Content

There has been growing concerns worldwide about the environmental impacts of conventional cropping systems. For an agricultural system to be sustainable, it must be able to hold out against the vagaries of climate and adapt itself to social changes. The more complex and vastly diverse an agricultural system is, the more resilient it will be to the sharp changes in climatic fluctuations.

Different perspectives and approaches to farming are required for different soil and climatic conditions. Each of these perspectives and approaches should be based on the sole objective of increasing biomass production to augment active organic matter. Active organic matter ensures secure habitat and food for beneficial soil organisms that promote soil structure and porosity, supply nutrients to plants and enhance the soil water holding capacity.

Several instances have demonstrated the possibility of restoring soil organic matter levels. Activities promoting the accumulation and supply of organic matter invariably build up the soil organic matter content. Using cover crops, refraining from burning residues and such other practices that reduce the soil organic matter decomposition rates (e.g., reduced and zero tillage), are typical approaches in this direction.

Increase in soil organic matter content is brought about by a set of factors that result in:

- Increased biomass yield;
- Increased organic matter stock;
- Decreased decomposition toll of soil organic matter.

19.2.1. Increased Biomass Yield

a. Water Availability to Crops

Ensuring adequate water availability to crops during dry conditions through irrigation or water harvesting measures enhances biomass yield, soil biological activity and plant residues that provide organic matter. Water harvesting measures include several technologies for runoff utilization and management. They range from capture and concentration of runoff (to be utilized by a discrete annual or perennial crop for enhanced growth and yield) to its storage to meet supplemental irrigation and domestic and livestock requirements.

Devising a water harvesting system guarantees that the water harvested during runoff periods is stored either directly in the soil for prospective crop use or stored in collection ponds or small farm reservoirs. Thus, crop production is stabilized through assured availability of soil moisture and farm profits are augmented further by stretching crop production into the dry season.

The factors to be reckoned with respect to the runoff farming systems and reservoirs include site characteristics, watershed size, rainfall distribution, quantum of expected runoff and water requirements of crops. If, in a reservoir, a minimum water depth of about 1 metre can be maintained, fish culture can provide additional income to the farm.

b. Balanced Fertilization

Where there is ample supply of nutrients in a soil, crops grow well and yield good amounts of biomass. Fertilizers are imperative as supplementary nourishment to crops in those cases where the soil lacks in fertility and, therefore, cannot produce optimum biomass yields. However, fertilizers should be applied in right quantities, in right proportions, in right chemical forms and by right methods. Unbalanced fertilization may lead to unhealthy plants. Especially with N, unbalanced fertilization may result in higher weed population, higher pest incidence and poor produce quality.

Where the organic matter content of the soil is high, the fertilizer use efficiency of the crops will be high. In very poor or depleted soils, the fertilizer use efficiency of crops is also poor. When the soil organic matter levels are restored, fertilizers maintain the revolving buffer of nutrients in the soil by increasing crop yields and, hence, the quantity of residues returned to the soil.

c. Cover Crops

Raising cover crops on farm land is one of the best practices to enhance soil organic matter levels. By growing cover crops, the following benefits accrue:

- Cover crops prevent erosion by binding soil and lessening the impact of raindrops.
- They replenish soil organic matter by adding plant material to soil.
- They bind excess nutrients in the soil.
- Some cover crops (e.g., rye) prevent leaching of nutrients.
- Leguminous cover crops fix nitrogen in the soil for future crop use.
- Most cover crops provide habitat for beneficial insects and other organisms.
- They protect soil organisms by abating soil temperatures.

An array of grains, legumes and oilseed crops have the potential to provide great benefits to the soil and can be used as vegetative cover. However, some crops bring in certain additional benefits which should be taken cognizance of when a rotation scheme is planned. In the first years, it is prudent to begin with crops that potentially cover the soil surface with bulks of slow-decomposing residues (because of the high C:N ratio). Grasses and cereals are most pertinent here. Their intensive rooting systems improve the soil structure.

In the years following initial years, when soil health shows definite signs of improvement, leguminous crops that potentially enrich the soil with nitrogen and whose residues decompose readily because of their low C:N ratio can be incorporated in the rotation. Later, when it is fairly certain that the system is stabilized, economically significant cover crops (such as livestock fodder) can be included.

While selecting cover crops, crops with high lignin and phenolic acid contents should be considered because they give the residues a greater resistance to decomposition. This results in the protection of soil for longer periods and the humus produced is far more stable.

Another factor to be reckoned while selecting cover crops is the biochemical composition of the residues. Depending on the species, their chemical composition and the kind of management practices adopted, there are bound to be differences in decomposition rates of residues. The cereals (such as oats and wheat) are relatively more resistant to decomposition than legumes. The legumes, by virtue of their low C:N ratio and low lignin content, undergo a rapid decomposition. Farm production systems in which residues are left on the soil surface (such as direct seeding and the use of cover crops) stimulate the proliferation and activity of soil faunae many folds.

A crop in the rotation that is grown with the sole objective of incorporating fresh vegetative matter in the soil is specifically referred to as *green manure* crop. Though a green manure crop is raised primarily to add organic matter to soil, this is not the best way of using of organic matter, more so in hot climates, for two reasons:

- When a green manure crop is incorporated in the soil all at once, there is a short period of intense microbial activity in the decomposing the material. This hype in microbial activity leads to a sudden spurt in the release of nutrients that the seedlings of the following crop are not able to utilize fully and the nutrients are thus lost from the system.
- While growing green manure crop, mechanical disturbance of the soil, though inevitable, should be kept to the barest minimum possible.

In general, if the production of green manure biomass is greater, the population of the soil faunae will also be greater. The dynamics of surface residue decomposition depend, among other things, on the activity of soil faunae. The integration of the residues into the soil is facilitated by the soil macrofaunae - chiefly, earthworms, beetles, termites, ants, millipedes, spiders, snails and slugs. Soil structure, porosity and water percolation are vastly improved through burrows, ingestions and secretions.

The process of incorporation of cover crops and weed residues from the soil surface deep into the subsurface layers by soil macrofaunae is naturally slow. The macrofaunae provide microorganisms with food and air through bioturbation and regulate their activities. In this way, plant nutrients are slowly and steadily released and supplied to the crops for prolonged durations. Simultaneously, the residues cover the soil for a longer period of time and protect it from the adverse impacts of rain and the Sun.

d. Dense Vegetation

Low plant densities limit crop yields. A wider plant spacing is often intended to give 'some rest' to the soil; but, in real terms, it indicates the impoverishment of the soil. Farmers usually determine plant spacing duly considering soil fertility and water availability. They often space plants widely on depleted soils in arid and semi-arid regions so as to ensure an adequate availability of plant nutrients and water for all plants.

However, it is imperative to adopt the recommended plant spacing to ensure most optimum rooting density and biomass production. This conducive ambience guarantees the availability of organic matter for food and habitat for soil organisms. In an established crop, subdued sunlight in between closer crop rows may hamper regrowth of weeds.

Crop yields are increased by the planting pits in severely degraded lands under semi-arid conditions. Whereas rainfall is collected near the plants, soil faunal activity and organic matter accumulation are concentrated in the planting pits. Planting pits have been a successful conservation practice for small-holding Zambian peasants who do not have availability of sufficient fertilizers or tractor services.

e. Agroforestry Systems

By *Agroforestry,* we refer collectively to different land-use systems in which woody perennials (trees, shrubs, etc.) are integrated in the farming systems. Agroforestry encompasses a wide range of farming systems in which food crops, forestry and pasture species are combined in multiple ways (agrosilviculture, silvipasture and multipurpose forest production). Generally, there are two different approaches to agroforestry. In the first, agricultural crops and pasture are incorporated as a transitional means of utilizing the land until forest plantations are fully established. In the second, trees and shrubs are integrated permanently into the crop and animal production systems for the twin benefit of land resource protection and crop production.

Alley cropping is an agroforestry system of farming in which annual crops are raised between rows of planted trees or woody shrubs. These trees and woody shrubs are pruned appropriately during the cropping season to furnish green manure and to minimize shading of crops grown in tandem. In alley cropping, short trees or shrubs are planted in rows in croplands, mostly along the contour (Fig. 19.3). Depending on the slope, soil type and its susceptibility to erosion, crop species and rainfall, the spacing between rows is decided.

Fig. 19.3: Alley cropping of cotton in-between pines

Perennial trees and shrubs in alley cropping provide (by means of litter and pruning) organic matter to the system and recycle plant nutrients from subsoil layers through their root network. The finest contribution of perennial trees and shrubs is that all through the year their roots excrete root exudates and the decaying root cells serve as energy source for soil microbes. Thus, the food web in the soil is sustained, even in the context when annual crops are not raised during dry seasons. The ultimate result is that soil biota is in place, ready to provide nutrients to the crop at the start of the ensuing cropping season.

The easiest and cheapest possible way to establish hedgerows is by direct seeding. The emerging seedlings should be carefully nurtured to survive competition from weeds. Raising saplings in a nursery and transplanting them in the field may be obligatory for some species. Other species may be established by cuttings. If good establishment is ensured, the plants will be able to fend off both dry spells and browsing livestock. In a successfully established hedgerow, the selected plants will be tall and outgrow the weeds at the time of harvest of the first crop.

Crop production can be impacted by the hedgerow species competing for light, water and nutrients and due to other complex factors like allelopathy, pests, diseases, etc., that are unique to the agroforestry systems. Therefore, the hedgerow species will have to be chosen after careful consideration of the above-mentioned complexities.

During the cropping season, hedgerows may have to be pruned to avoid shading of the crop. The timing, frequency and extent of pruning may all depend on the species and the season. As a general rule, pruning is required less frequently if the hedgerows are shorter and the crops are taller. Fast-growing species such as *Gliricidia sepium* and *Leucaena leucocephala* may require pruning to a height of about 50 cm approximately every six weeks during the cropping season. Too frequent pruning could result in tree dieback, warranting cautious approach.

Integrating trees and woody shrubs into the cropping system offers several additional benefits. Many farmers may not be inclined to accept these benefits, if their positive effects are not discernible. Also, farmers who are on short-term land tenure may not evince interest in these benefits. As agroforestry systems are not amenable for mechanization, they incur high labour inputs, more so for hedgerow pruning. Where the grazing livestock are allowed free access, establishment of hedgerows can be a difficult proposition, if adequate measures are not taken to protect the young plants.

Increased labour inputs, reduced cropped area and inherent difficulties in mechanization may render alley cropping uneconomic, unless the hedgerow species incorporated in the system yield direct benefits such as fruits, fuelwood or poles/timber for construction purposes.

f. Afforestation

Afforestation is the establishment of tree species on a land that has not grown trees recently. It serves the twin purposes of protecting erosion-prone areas and rehabilitation of degraded land. Afforestation especially provides protective cover in vulnerable, steep and mountainous areas. It provides fuelwood and fodder and also replenishes timber resources.

The establishment of a forest cover is an effective way of enhancing organic matter production. The choice of species and the selection of an appropriate site are important for successful afforestation which differs according to climate, soil, slope and the intended purpose of the forest (whether it is for timber production or livestock grazing, etc.). The land chosen must support an appropriate forest type.

The first requirements in any afforestation initiative are adequate quantities of good quality seed of the chosen species or provenances. The seeds of several tree species require special pre-treatment for good germination and uniform crop stand. The pre-treatments may vary in their nature and duration: soaking the seed in water for varying periods; alternatively soaking in water and drying; scarifying the seed to render its seed coat permeable to water; dropping the seed into boiling water, or, boiling it for a short time. Saplings of some tree species need mycorrhizal inoculation if they are planned for soils devoid of mycorrhizae (e.g., *Casuarina* in sandy soils). The aim is to ensure rapid and uniformly good germination.

Afforestation can be achieved by direct field sowing of seeds or planting saplings from a nursery. Direct sowing of seeds is definitely cheap, but is usually much less reliable; it is justified only in situations where:

- the seed is not in dearth, but bountiful;
- adequate germination can be ensured under field conditions;
- the seedlings are able to circumvent any adverse climatic condition in the immediate period following germination;
- the growth rate is fast enough, obliterating the need for a prolonged period of tending and weeding.

g. Regeneration of Forests and Grasslands

Regeneration of natural vegetation (such as forests and grasslands) enhances biomass yield and species diversity. This leads to plenty of diverse soil biota and co-habited beneficial organisms. Natural vegetation offers multiple benefits. Besides supplying fuelwood, fibre, biocontrol (neem) and medicinal products and providing habitats for various beneficial species (pollinators and natural enemies) and wildlife, it restores soil fertility (*Acacia albida* and other leguminous species).

19.2.2. Increased Organic Matter Stock

a. Safeguard from Fire

Protection of organic materials from fire stabilizes the soil organic matter stock significantly. Fire destroys all organic materials on the land surface and the damage depends on the intensity of fire which is decided by the vegetation type and climate. Fire sterilizes the surface soil which loses its organic matter; the population of soil micro- and macrofaunae is reduced and ready-to-use organic matter is not available for quick restoration of the populations. Burning also significantly affects organic matter recycling. However, in some countries like Africa, burning is widely practised to stimulate pasture regrowth for livestock and to control pests and diseases.

b. Residue Management

In crop production systems where crop residues are well-managed, they:

- improve the tilth of the seedbed;
- enhance the soil moisture content by capturing rainfall;
- enhance soil water infiltration and retention capacity;
- reduce evaporation from the soil surface;
- sequester C in the soil;
- buffer the pH by adding soil organic matter;
- facilitate nutrient availability for soil biological activity and plant growth;
- protect the soil from erosion.

Whether the residues should be distributed evenly over the field or left intact is decided depending on the nature of the succeeding crop. An even distribution of residues (i) provides homogenous temperature and humidity conditions at the time of sowing, (ii) facilitates even sowing, uniform germination and emergence of seedlings, (iii) minimizes pests and diseases and (iv) inhibits weed population through allelopathic effects.

The ideal method of managing crop residues depends on the purpose of crop residues and on the equipment available with the farmer. If the desired purpose is to maintain a mulch over the soil for as long as possible, the best way of managing biomass is by breaking it down, but without killing it, using an appropriate tool such as knife roller. If the desired purpose is to commence the decomposition process immediately for release of nutrients, the residues should be mown or slashed, followed by some N application (to lower the C:N ratio, because dry residues have a high C:N ratio).

c. Forage Grazing

A significant loss of soil organic matter occurs when crop residues are removed from the field. In situations where animal manure is not returned to the field, this loss would be considerable. If controlled grazing is allowed in a field, the return of animal manure to the field can be ensured without much of labour input. Crop and livestock components can be successfully integrated with judicious allocation of crop residues between the two. The quantum of residues produced by the system is sufficient to meet the requirements both as soil cover and as livestock fodder, especially by using local varieties (instead of high yielding varieties) with a high straw/grain ratio.

In northern Tanzania, farmers segregate the palatable and non-palatable parts of the crop residues. They use the non-palatable parts for covering the soil to serve as food for soil organisms and feed the palatable parts to cattle and goats they keep in proximity to the homestead. Though labor-intensive, this practice is a judicious compromise between using the residues for grazing or for soil cover.

d. Pest Management

Prudent on-field pest and disease management practices can ensure healthy crops. And healthy crops will produce optimal biomass essential for building up soil organic matter. Diversified cropping and mixed crop-livestock systems ensure biological control of pests and diseases *via* species interactions. By providing a healthy environment for their crops through integrated pest management, farmers learn to oversee their crops regularly for pests, natural enemies and cases of damage. On that score, they decide whether to use locally available natural products (such as neem) or to go for use of pesticides.

e. Organic Wastes Application

Application of animal manure, slurry or other organic wastes increases the organic matter content of the soil. In some cases, it is ideal to let a period of decomposition before field application. Immediately upon application, carbon-rich wastes temporarily immobilize available N in the soil due to the action of microbes that need both C and N for their growth and proliferation. Nitrogen immobilization is minimal with animal manure which is usually rich in N. In case straw is a part of the manure, its application after a decomposition period avoids N immobilization in the field.

f. Composting

Composting is a technology advocated for recycling organic materials in a useful way for achieving enhanced agricultural production. Composting converts organic materials into a more stable humus for soil application through biochemical processes that accelerate the rate of decomposition under controlled conditions in compost heaps and pits.

Compost heaps are best suited to humid environments where watering the compost is possible. Compost pits should have a minimum capacity of 1 m^3, but not be deeper than 70 cm, and should be underlain with some rough material to ensure good aeration. In dry composting, the compost is covered with soil to create an anaerobic environment. But this is a slower process as compared to the more usual moist, aerobic process. Since the C:N ratio of the compost pile is a criterion for optimal microbial activity, an admixture of soft, green and tough, brown materials is used. To accelerate the composting process, ash and phosphate rock are often times added as supplements.

Compost can complement certain crop rotations and agroforestry systems. Akin to soil organic matter in composition, compost breaks down slowly in the soil (in contrast to manure and sludge which decompose relatively fast, yielding a glut of nutrients essential for crop growth) and vastly improves the soil physical conditions. In many instances, it takes much longer time to rejuvenate a poor soil using these practices because the amount of organic material that is added is trivial in proportion to the mineral content of the soil.

Composting, as a technology, is sustainable depending on the adequate availability of organic materials, water, manure and 'affordable' labour. In places where the availability of these inputs is vouched for, composting can play a significant role in productive agriculture. Compost exerts ameliorative effects on soil physical, chemical and biological properties. Well-made out compost is characterized by the presence and availability of all the nutrients needed by crop plants and can potentially improve soil fertility as well as regenerate degraded soil. However, materials for compost production may often be in short supply and the technology warrants high labour inputs. Therefore, more often than not, compost application may have restricted application areas, such as growing vegetable in home gardens.

g. Mulching

One of the effective ways to mend a soil needing amelioration is to mulch the land. Mulches are protective materials placed on the surface of the soil to safeguard it from the impact of rains and erosion and to enhance its fertility. Crop residue mulching is a system of maintaining a protective cover of crop residues (straw, stalks, stubbles, palm fronds, etc.) on the soil surface that is especially precious where a satisfactory plant cover cannot be established in quick time and where the risk of soil erosion is greatest.

Mulching adds organic matter to the soil, reduces weed growth and virtually eliminates erosion during the entire period when the land is covered with mulch. There are two types of mulching systems:

- *In situ* mulching: Plant residues are left to remain at the very same place where they fall on the ground.

- Cut-and-carry mulching: Plant residues are far-fetched from somewhere else and used as mulch.

Crop residue mulching has several positive impacts on crop production. However, it might be that the existing cropping practices have to be changed. To cite an example, farmers may traditionally burn crop residues *in lieu* of returning them to the soil. The utility of *in situ* mulching hinges on designing appropriate cropping systems and crop rotations that are to be integrated with the farming system. The labour-intensiveness of cut-and-carry systems presents an impeding constraint. Mulch may be more pertinent and useful to home-gardens and valuable horticulture crops than in less intensive cropping systems.

Mulch influences the soil biota a lot. Placement of litter on the soil surface (as against incorporation by ploughing) increases the ratio of fungi to bacteria, because the fungi assimilate carbon more efficiently than the bacteria. Additionally, it encourages bioturbation (mixing) by macrofaunae that drag the materials into surface layers of the soil. Zero tillage with medium to high amounts of stover mulching increased soil organic carbon stocks in the upper 0 - 40 cm of soil by > 0.4% per year in a Mollisol of north-eastern China.

19.2.3. Decreased Decomposition Toll of Soil Organic Matter

Zero Tillage

Conventional repetitive tillage breaks up aggregates and destroys soil structure. It reduces the soil's potential to hold moisture, scales down the soil organic matter content and along with it the population of soil faunae (e.g., earthworms) that play a significant part in nutrient cycling and soil structure.

Evading mechanical disturbance of soil entails keeping the soil undisturbed since the harvest of the previous crop and growing crops without mechanical seedbed preparation. This practice, termed *zero tillage*, is used synonymously with the terms such as no-tillage, no-till farming, direct seeding and direct drilling.

Conventional tillage increases mineralization rate by stimulating the heterotrophic microbial activity through soil aeration. By destroying soil structure, it also restricts the movements of soil faunae (such as earthworms), which facilitate humus production through the ingestion of fresh residues.

Reduced or zero tillage has two distinct benefits over conventional tillage *vis-à-vis* soil organic matter. Firstly, it regulates the activity of heterotrophic microorganisms because the pore atmosphere is richer in CO_2/O_2 and secondly, it promotes the activity of the 'humifiers'.

Tillage is most commonly practised to control weeds, but mulching is environmentally a superior practice for weed control. The loosened soil resulting from tillage has less structure than before. This loosened soil is subsequently packed by traffic or heavy rain, negating the expensive tilling that produced the loose soil and culminating in a poor ambience for water infiltration, seed germination and root growth. Tilling is again required to re-loosen the soil entailing additional expenditure with the same outcome: subsequent repacking and downgraded soil structure. This is a typical downward gyring of conventional agriculture. Also, when a moist soil is tilled, it leads to compaction and if a dry soil is tilled, it leads to pulverization of the soil.

Widespread adoption of no- or zero-tillage systems for cropping has stemmed from accelerated soil erosion and the high input costs that are inherent with plough-based conventional methods of cultivation. Studies have shown that no-till system has the potential to decrease overall production cost, while decreasing the mean soil loss by 86% as compared to the conventional system.

In no-tillage systems, the crop is sown in a land that has not been disturbed since the harvest of the last crop. Crop residue mulch is kept firmly secured to the ground. Weed control is done either mechanically or by use of cover crops. In some cases, contact herbicides may have to be used.

In zero-tillage systems, soil faunae gradually restart their pedoturbation activities leading to the loosening of the soil and mixing up of the soil components. This is termed, '*biotillage*', the extra benefit of which is the creation of a firm, porous soil structure without expensive, potentially degenerating and time-exhausting cultivations.

In zero-tillage systems, the soil macrofaunae slowly integrate cover crops and weed residues deeper down into the soil from the soil surface. They also provide the microorganisms with food and air facilitated through their burrows. In this way, they regulate the microbial activity and nutrients are released gradually to the succeeding crop.

Some crop rotations and zero tillage have been shown to support *Bradyrhizobia* populations, nodulation and nitrogen fixation. An increase in population size of root nodule bacteria, as high as 200 to 300 per cent over conventional tillage has been reported in a zero-tillage system. The inclusion of soybean in the crop rotation resulted in a five- to ten-fold increase in population size of the same bacteria as compared to cropping systems devoid of soybean.

By definition, the term zero tillage applies to methods wherein the soil is not at all disturbed whatsoever. Broadcasting of seeds over the previous crop residues is a way of adopting zero tillage. If needed, the residues are shaken to ensure that the broadcast seeds fall on the soil surface.

Direct drilling is a method of sowing seeds such as wheat, barley, maize, sorghum and soybean directly into the shallow furrows made into the previous crop residues. Here, weeds can be controlled mechanically by knocking them down and breaking their stems with a knife, or chemically with herbicides.

Traditional practices, such as the burning of crop residues, may not be amenable for no-tillage systems. Many conflicting situations may arise whether to leave crop residues on the soil surface or to feed them to livestock, especially in seasons when there is a shortage of fodder.

In intensely mechanized agriculture, soil compaction is brought about through the impact of wheels of heavy machinery (Fig. 19.4). Under zero-tillage systems, care should be taken to reduce both the random placement of wheels in the fields as well as the compaction from animal hooves.

Fig. 19.4: Soil compaction is caused by wheel impact of machinery

Contrary to the general belief that draught animals cause lesser land degradation than tractors, there are reports of soil compaction brought about by the action of draught animals on small-holding farm enterprises in Bangladesh and Malawi. The hooves of draught animals and the shearing effect of ploughs when repeatedly used at the same depth can cause acutely compacted layers. The risk of soil compaction is greatest when grazing animals are let in zero-tillage fields in moist or wet soil situations.

To avoid the problems associated with heavy machinery, the wheels of all equipment are made to follow permanent, defined tracks in a field in what is known as '*controlled traffic*'. This ensures that compaction is confined to certain specifically known areas. Otherwise, all large tractors can be fitted with flotation tyres to reduce the soil compaction risks.

Recent research has highlighted the detrimental effects of soil compaction from the impact of machinery wheels on the survival of earthworms. Earthworms were far greater in number under controlled traffic than under wheeled traffic. Though there were indications that earthworms can survive an initial compaction in the field for a short span, where wheeling was followed by tillage, the earthworms' survival rate was much better. Where it was not, earthworms were immobilized and disadvantaged a lot.

20

Analytical Procedures

Methods and procedures for the qualitative characterization and quantitative analysis of different constituents of soil organic matter are presented in this Chapter.

20.1. Moisture

The soil sample should be in an air-dry condition after mixing, sieving and grinding. When stored in a moisture-tight container, the free H_2O content should not vary appreciably during storage. While most determinations can be done on air-dry samples, for certain purposes such as obtaining summation values, a moisture content value is required.

Weigh 2 g soil into a tarred, shallow, stoppered weighing bottle and heat with the bottle uncovered for several hours, or overnight, in a drying oven at 110 °C. Cover, cool in a desiccator and weigh. Report the percentage loss in weight as moisture. The dried sample may be held for determination of loss on ignition.

20.2. Organic Matter

The most accurate procedure for the determination of organic matter in soils is the dry-combustion method for total carbon in which the CO_2 evolved on heating is trapped in some absorbent and a correction is made for any native CO_3^{2-} present. Since this is a tedious method, it is often felt that one of the faster, wet-oxidation procedures using chromic acid will give results of a desired accuracy. As an approximate method, loss of weight of an oven-dry sample on ignition is a method often employed.

20.2.1. Loss on Ignition [As modified from Ben-Dor & Banin, 1989]

Heat crucibles in a muffle furnace at 400 °C for 2 hours, cool and determine the tare weight to 0.1 mg. Add 1 to 3 g of air-dried soil ground to < 0.4 mm to a tarred crucible and heat at 105 °C for 24 hr. Cool the crucible in a desiccator over $CaCl_2$ and determine the weight of crucible plus sample to 0.1 mg. Obtain the weight of over-dried sample by subtraction. Ignite samples in a muffle furnace at 400 °C for 4 hour. Cool the crucible in a desiccator over $CaCl_2$ and determine the weight of crucible plus ignited sample to 0.1 mg. Calculate the weight of ignited sample by subtraction. Report the percentage loss in weight as "loss on ignition". The organic matter is assumed to equal loss on ignition in most surface soils.

$$\text{Loss on ignition (\%)} = \frac{\text{Weight}_{105} - \text{Weight}_{400}}{\text{Weight}_{105}} \times 100$$

Weight_{105} = Weight of soil sample after heating at 105 °C

Weight_{400} = Weight of soil sample after ignition at 400 °C

Key Reference

Ghimire, R., Shah, S. C., Dahal, K. R., Shrestha, A. K., Adhikari, C., Lauren, J. G. and Duxbury J. M. 2007. Equation for predicting soil organic carbon using loss on ignition for Chitwan valley soils. *IAAS Res. Adv.* 1: 229- 232.

20.2.2. Volumetric Estimation of Soil Organic Matter [Walkley & Black, 1934]

Reagents

1. Concentrated sulphuric acid
2. 1 *N* Potassium dichromate [Dissolve 49.04 g $K_2Cr_2O_7$ in 1000 mL of distilled water]
3. Orthophosphoric acid (85%)
4. Diphenyl amine indicator [Dissolve 0.5 g of diphenyl amine in 100 mL of concentrated H_2SO_4 acid]
5. 0.5 *N* ferrous ammonium sulphate [Dissolve 392.0 g of $FeSO_4.(NH_4)_2SO_4.6H_2O$ in distilled water. Add 15 mL of concentrated H_2SO_4 and make up to 2000 mL with distilled water]

Procedure

Weigh accurately 1.0 g of soil ground to pass through a 0.2 mm sieve and transfer it to a 500- mL Erlenmeyer flask. Add 10 mL of 1.0 *N* $K_2Cr_2O_7$ and 20 mL of concentrated H_2SO_4 and swirl the contents a few times. Allow the flask to stand for 30 minutes and then add 200 mL of distilled water, followed by 10 mL of orthophosphoric acid and 1 mL of diphenyl amine indicator. Titrate the contents with 0.5 *N* ferrous ammonium sulphate till the color flashes from blue violet to green. Simultaneously run a blank without soil.

Calculations

Weight of soil taken = W g

Volume of 0.5 *N* ferrous ammonium sulphate solution used in blank = V_1 mL
Volume of 0.5 *N* ferrous ammonium sulphate solution used in back titration = V_2 mL

Volume of 0.5 *N* ferrous ammonium sulphate solution used for soil = ½ (V_1 – V_2) mL

1 mL of 1 *N* $K_2Cr_2O_7$ = 0.003 g of organic carbon

½ (V_1 – V_2) mL of 1 *N* $K_2Cr_2O_7$ = [0.003 x ½ (V_1 – V_2)] g

Per cent of organic C in soil = [0.003 x ½ (V_1 – V_2)] x [100 / W] %

Per cent of organic matter in the given soil = Organic C % x 1.724

20.2.3. Colorimetric Determination of Soil Organic Matter[9]

The oxidation of organic materials by dichromate in acid media involves the half reaction: $Cr_2O_7^{-2} + 14H^+ + 6e^- \rightarrow 2Cr^{3+} + 7H_2O$. Thus, appearance of Cr^{3+} in the reaction media of dichromate oxidations of organic matter should be in direct proportion to the quantity of organic matter oxidized. Analysis of the reaction media would then provide an index for organic matter content of the sample. The Cr^{3+} exists in acid aqueous media as the hexaquo ion, $[Cr(H_2O)_6]^{3+}$, which is very stable. Chromium (III) has two broad absorption maxima in the visible range, one near 450 mμ and the other near 600 mμ. The dichromate ion also has absorption maximum near 450 mμ, but does not absorb near 600 mμ. Thus, oxidation of organic matter with dichromate followed by Cr^{3+} absorption near 600 mμ provides a satisfactory method of soil organic matter determination.

Reagents

1. Concentrated sulphuric acid
2. 1 *N* Potassium dichromate [Dissolve 49.04 g $K_2Cr_2O_7$ in 1000 mL of distilled water]
3. Sucrose (AR)

Procedure

Weigh accurately 1.0 g of soil ground to pass through a 0.2 mm sieve and

[9]*Adapted from:* Sims, J. R. and Haby, V. A. 1971. Simplified colorimetric determination of soil organic matter. *Soil Science* 112(2): 137-141.

transfer it to a 250-mL Erlenmeyer flask. Add 10 mL of 1.0 *N* $K_2Cr_2O_7$ followed by 20 mL of concentrated H_2SO_4. Swirl the contents of the flask and set aside for 20 minutes. Add 100 mL of distilled water, filter[10] and measure light absorption of the filtrate at 600 mμ. For soils containing more than 7 per cent OM, use 0.5 g samples and for soils containing less than 1.5% OM, use 2.0 g samples. Construct a calibration curve for determining the soil organic matter content.

Calibration

Prepare sucrose samples such that, without oxidation and at 100 mL volume, they would have 0, 2.92 x 10^{-3}, 5.84 x 10^{-3}, 8.76 x 10^{-3}, 11.68 x 10^{-3}, 14.60 x 10^{-3}, 17.52 x 10^{-3} and 20.44 x 10^{-3} *M* concentration. These concentrations are equivalent to organic matter contents of 0, 1, 2, 3, 4, 5, 6 and 7 per cent in 1 g soil samples. After oxidation is complete, determine light absorption at 600 mμ using sucrose-free sample as the reference solution.

20.2.4. Fractionation of Soil Organic Matter Based on Solubility Characteristics[11]

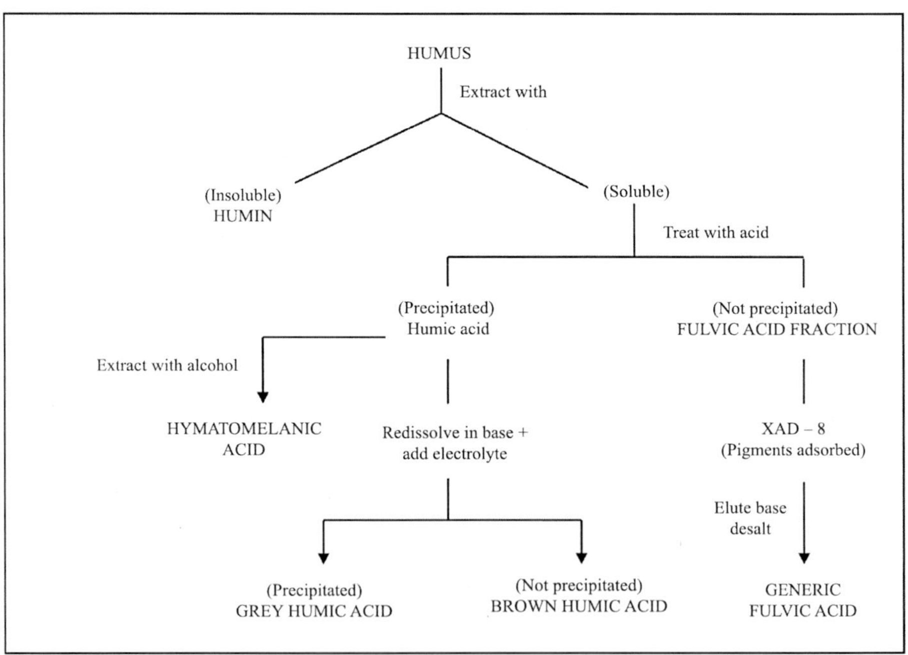

---

[10]Filtration through filter paper may result in erroneous high values for organic matter content; instead, filtration through sintered glass is recommended.

[11]*Adapted from*: Stevenson, F.J. 1994. *Humus Chemistry. Genesis, Composition, Reactions*. John Wiley & Sons Inc., NY. p. 41.

Reagents

1. 0.1 *N* HCl and 1.0 *N* HCl
2. 0.5 *N* NaOH
3. Concentrated HCl acid

Apparatus

1. Centrifuge and centrifuge bottles
2. Glass wool and pH indicators

Procedure

Acid Washing: Equilibrate the sample with 0.1 *N* HCl at room temperature. Adjust the solution volume to provide a final concentration that has a ratio of 10 mL liquid per 1 g dry sample. Shake the suspension for 1 hour and then separate the supernatant from the residue by decantation after allowing the solution to settle or by low-speed centrifugation.

Place a 40-g sample of acid washed (0.1 *N* HCl) soil (20 g for peat and organic-rich samples) into a 250 mL polythene centrifuge bottle, add 200 mL of 0.5 *N* NaOH solution and stopper the bottle tightly with a rubber stopper. Shake the mixture for 12 hours on a mechanical shaker, wash down the sides of the bottle with distilled water and centrifuge the mixture. Decant off dark-colored supernatant liquid, filter it through glass wool to remove suspended material and adjust the pH of the solution to about 1.0 with conc. HCl. [All steps should be carried out under a blanket of N_2 gas so as to minimize chemical changes in the extracted organic matter.]

Add an additional 200 mL of 0.5 *N* NaOH solution to the soil, shake the mixture for 1 hour and repeat the centrifuging and decanting procedures. Disperse the residue in 200 mL of distilled water, centrifuge the mixture and add supernatant liquor to the previous extracts. Adjust the pH of the resulting solution to 1.0 with conc. HCl and allow the humic acid to settle down. If a large quantity of soluble organic matter is required, repeat the extraction procedure with fresh sample of soil.

Siphon off the excess supernatant liquor (fulvic acid) from the acidified extract, transfer the remainder of the suspension to an 8-oz polythene bottle and centrifuge off the humic acid. Redissolve the humic acid in 0.5 *N* NaOH solution, reprecipitate it by adjusting the pH of the solution to 1.0 with concentrated HCl and then centrifuge out the humic acid. In each case, add the supernatant liquor to the original acid filtrate. Repeat the purification procedure a second time and then wash the humic acid precipitate several times with small quantities of H_2O to remove the residual HCl. Dry the humic acid (preferably by freeze-drying) and grind it to a brown powder.

20.2.5. The E_4/E_6 Ratio

The ratio of absorbances at 465 and 665 nm, referred to as the E_4/E_6 ratio, has been widely used for characterization purposes. The E_4/E_6 ratios for humic acids are usually <5.0; those for fulvic acids range from 6.0 to 8.5.

Low E_4/E_6 ratio means	High E_4/E_6 ratio means
High molecular weight	Low molecular weight
High condensation	Less condensation
Aromatic nature	Aliphatic nature
Highest mean residence time	Lowest mean residence time
Points to older material	Relatively young material

The best procedure to determine the E_4/E_6 ratio is to dissolve 2 to 4 mg of the humic or fulvic acid in 10 mL of 0.05 *N* $NaHCO_3$, which gives an optimum pH for absorbance measurements (wavelengths of 465 and 665 nm).

Procedure

Place a 40-g sample of acid washed (0.1 *N* HCl) soil (20 g for peat and organic-rich samples into an 8 oz polythene centrifuge bottle, add 200 mL of 0.5 *N* NaOH solution and stopper the bottle tightly with a rubber stopper. Shake the mixture for 12 hours on a mechanical shaker and centrifuge the mixture. Decant off dark-colored supernatant liquid and measure the absorbances at 465 and 665 nm. Determine the E_4/E_6 ratio.

20.3. Humic Acid Analysis

20.3.1. Total Acidity

Total acidity = COOH plus Phenolic OH and/or Enolic OH. The sample is allowed to react with excess $Ba(OH)_2$ and the unused base is titrated with standard acid:

$$2HA + Ba(OH)_2 \rightarrow BaA_2 + 2H_2O$$

To between 50 mg and 100 mg of humic preparation in 125-mL glass-stoppered Erlenmeyer flask, add 20 mL of 0.2*N* $Ba(OH)_2$, stopper and shake for 24 hours at room temperature. Run a blank simultaneously. Filter the suspension and wash residue thoroughly with CO_2-free distilled water and titrate the filtrate plus washings potentiometrically (glass-calomel electrodes) with standard 0.5*N* HCl solution to pH 8.4.

$$\text{Total acidity (meq g}^{-1}) = \frac{(V_b - V_s) \times N \times 10^3}{\text{mg of sample}}$$

Where V_b and V_s represent the volumes of standard acid used for the blank and sample, respectively and N is the normality of the acid.

20.3.2. Carboxyl Groups

To between 50 mg and 100 mg of humic preparation in 125-mL glass-stoppered Erlenmeyer flask, add 10 mL of 1*N* $Ca(CH_3COO)2$ and 40 mL of CO_2-free distilled water, stopper and shake for 24 hours at room temperature. Run a blank simultaneously. Filter the suspension, wash residue thoroughly with CO_2-free distilled water, combine the filtrate and washing and titrate potentiometrically (glass-calomel electrodes) with standard 0.1*N* NaOH solution to pH 9.8.

$$\text{COOH (meq g}^{-1}) = \frac{(V_b - V_s) \times N \times 10^3}{\text{mg of sample}}$$

Where V_b and V_s represent the volumes of standard base used for the blank and sample, respectively and N is the normality of the base.

20.3.3. Phenolic OH

The quantity of phenolic OH (more correctly, acidic OH) (in meq g^{-1}) is calculated as the difference between total acidity (in meq g^{-1}) and COOH content (in meq g^{-1})

Phenolic OH = Total acidity – COOH

20.4. Organic Nitrogen

Potentially Available Nitrogen by Alkaline Extraction[12]

The term 'potentially available nitrogen' includes nitrogen hydrolyzed from soil organic matter and small amounts of exchangeable ammonium that may be present in the soil.

Procedure

Extract 10 g of soil by shaking for 30 minutes with 50 mL of 0.25 *N* $NaHCO_3$ + 0.25 *N* Na_2CO_3 solutions. Centrifuge the suspension and digest the extract with 5 mL of conc. H_2SO_4 for 1 hr. Estimate the ammonia evolved on distilling the digest with 25 mL of 40% NaOH by the usual procedure. Express the result in kg ha^{-1}.

[12]Prasad, R. 1965. Determination of potentially available nitrogen in soils - A rapid procedure. *Plant and Soil* 23 (2): 261-264.

20.5. Organic Forms of Nitrogen[13]

Earlier methods used to characterize the organic-N compounds in soils were essentially modifications of Van Slyke (1911-12, 1915) method of protein analysis. However, only 20 to 40% of the total-N in surface soils could be definitely accounted for by these methods as amino acid-N and that a significant amount of the N in soils is in the form of 2-amino sugars (hexosamines). Therefore, attention was directed toward development of hydrolytic methods of characterization which provide reliable estimates of the amounts of total-, ammonium-, amino acid-, and hexosamine-N liberated by acid hydrolysis of soils.

20.5.1. Preparation of Hydrolysate

Reagents

1. 6 *N* HCl
2. 0.5 *N* NaOH
3. Approximately 5 *N* NaOH
4. Approximately 10 *N* NaOH
5. H_2SO_4, concentrated
6. 0.005 *N* H_2SO_4
7. Potassium sulphate catalyst mixture: Prepare an intimate mixture of 100 g of K_2SO_4, 10 g of $CuSO_4.5H_2O$ and 1 g Se. Powder the reagents separately before mixing and grind the mixture in a mortar to powder the cake which forms during mixing.
8. MgO: Heat heavy MgO in an electric muffle furnace at 600 to 700 °C for 2 hours. Cool the product in a desiccator and store in a tightly stoppered bottle.
9. Boric acid-indicator solution: Dissolve 20 g of pure H_3BO_3 in about 700 mL of hot water and transfer the cooled solution to a 1-litre volumetric flask containing 200 mL of ethanol and 20 mL of mixed indicator solution prepared by dissolving 0.330 g of bromocresol green and 0.165 g of methyl red in 500 mL of ethanol. Then dilute the solution to volume with water and mix it thoroughly.

[13]Bremner, J. M. 1965. Organic forms of nitrogen. *In*: C. A. Black (Ed.) Methods of Soil Analysis, Part 2. Chemical and Microbiological Properties. *Agronomy Series No. 9*, American Society of Agronomy, Madison, Wisconsin, USA, p. 1238 – 1255.

10. Phosphate-borate buffer, pH 11.2: Place 100 g of $Na_3PO_4.12H_2O$, 25 g of $Na_2B_4O_7.10H_2O$ and about 900 mL of water in a 1-litre volumetric flask until the two salts are dissolved. Dilute the solution to 1 litre and store in a tightly stoppered bottle.

Apparatus

1. Steam distillation apparatus
2. pH meter or pH indicators

Procedure

Place a sample of finely ground (<100 mesh) soil containing about 10 mg of N in a 125 mL Erlenmeyer flask fitted with a ground-glass joint. Add 20 mL of 6 *N* HCl and swirl the flask until the acid is thoroughly mixed with the soil. Place the flask on an electric hot plate with a heat-control device and connect the flask to a Liebig condenser. Then pass cold water through the water jacket of the condenser and heat the soil-acid mixture so that it boils gently under reflux for 12 hours.

After completion of hydrolysis, allow the flask to cool and filter the hydrolysis mixture through a Buchner or sintered-glass funnel fitted with Whatman No. 50 filter paper. Use a suction filtration apparatus to collect the filtrate in a beaker. Wash the hydrolysis residue with small (5 to 10 mL) portions of water until about 60 mL of filtrate is collected in the beaker. Immerse the lower half of the beaker in crushed ice and neutralize the hydrolysate (pH 6.5 ± 0.1) by cautious addition of NaOH. Add the alkali slowly, and with constant stirring, to ensure that the hydrolysate does not become alkaline at any stage of the neutralization. Use 5 *N* NaOH to bring the pH of the hydrolysate to around 5, and complete the neutralization using 0.5 *N* NaOH. Transfer the neutralized hydrolysate to a 100 mL volumetric flask and dilute it to volume with washings obtained by rinsing the beaker. Then stopper the flask and invert it several times to mix the contents.

20.5.2. Total Nitrogen

Place 5 mL of the neutralized hydrolysate in a 50 mL distillation flask, add 0.5 g of K_2SO_4 catalyst mixture and 2 mL of conc. H_2SO_4 acid. Heat the flask cautiously until frothing ceases. Continue to heat until the mixture clears. Complete by boiling the mixture gently for 1 hour.

After completion of digestion, allow the flask to cool, add about 10 mL of water and shake. Cool the flask under cold-water tap and then steam-distil the contents with 10 mL of 10 *N* NaOH. Collect about 35 mL of distillate in boric acid-indicator solution and determine the N content in the distillate by titration with 0.005 *N* H_2SO_4 from a micro-burette. [1 mL of 0.005 *N* H_2SO_4 = 70 µg of N]

20.5.3. Ammonium N

Place 10 mL of the hydrolysate in a 100 mL distillation flask, add 0.07 ± 0.01 g of MgO and connect the flask to a steam distillation apparatus. Determine the amount of NH_3 liberated by steam distillation as described above. Stop the distillation when about 20 mL of distillate is collected. [Period of distillation, approximately 2 minutes]

20.5.4. [Ammonium + Hexsosamine]-N

Place 10 mL of the hydrolysate in a 100 mL distillation flask, add 10 mL of phosphate-borate buffer and connect the flask to a steam distillation apparatus. Determine the amount of NH_3 liberated by steam distillation as described above. Stop the distillation when about 20 mL of distillate is collected. [Period of distillation, approximately 4 minutes]

20.6. Amino-sugar Nitrogen[14]

Amino-sugar constitute a significant 5 - 12% of soil organic nitrogen and roughly 3% of soil organic carbon. Although up to 26 Amino-sugar have been found in microorganisms, the four Amino-sugar quantified in soil are glucosamine, galactoseamine, muramic acid and mannoseamine.

Researches have shown that among various organic fractions in the soil, concentrations of Amino-sugar N are highly correlated with responsiveness of soils to fertilizer N. In other words, accumulation of amino sugar N in soil reduces the yield response of crops to N fertilization. For soil amino sugar test, soil samples should be collected from a depth of 8 inches and sampling should be done 6-8 weeks after manure spreading or cover crop plow down to avoid any ammonium-N released in the soil by newly added organic materials. The results indicate organic N mineralization potential of the soil which is valid for 2-3 years. Therefore, when using this method, annual soil sampling is not needed.

A variety of different methods are available for determining aminosugars after hydrolysis of soil organic matter. Gas chromatographic methods are very sensitive and have a high specificity, but the derivatization procedure of the

[14]Indorf, C., Dyckmans, J., Khan, K. S. and Joergensen, R. G. 2011. Optimization of amino sugar quantification by HPLC in soil and plant hydrolysates. *Biol. Fertil. Soils* 47:387–396.

hydrolysis products into volatization components is time-consuming. High performance liquid chromatography also requires derivatization procedures to determine aminosugars; however, this can be carried out today automatically by the HPLC system as pre-column derivatization (with Ortho- phthaldialdehyde) using the sample injection device or as post-column derivatization (with 2-cyanoacetamide) using a second pump with reaction baths or coils. Appuhn *et al.* (2004)[15] were the first to describe a HPLC method to determine the four amino sugars simultaneously which was subsequently improved by Indorf *et al.* (2011).

Procedure

Mix sieved (<2 mm) and air-dried soil (400 mg) or oven-dried (40 °C) and steel ball-milled plant material (700 mg) with 10 ml of 6 *M* HCl. After 6 (soil) or 3-hour (plant material) hydrolysis at 105 °C, filter the samples over glass filters (Whatman GF/A). For the determination of the recovery rate, add 0.3 mL of a 150 µmol L^{-1} standard solution to 0.3 mL of a quartz sand hydrolysate. Evaporate a 0.3-mL aliquot to dryness at 40 - 45 °C to remove HCl, re-dissolve in water, evaporate a second time and re-dissolve in 1 mL water. To test the effect of sample pH on the amino sugar amounts, prepare another set of samples using 1 mL phosphate buffer solution pH 7 (containing 40.8 mmol L^{-1} $Na_2HPO_4.2H_2O$ and 25.9 mmol L^{-1} KH_2PO_4) to re-dissolve samples. After centrifugation at 5,000×g, freeze the supernatant and store at −18 °C until analysis.

Reagents

Prepare all solutions with Milli-Q water.

1. *Buffer solution (pH 11):* Dissolve 50 g H_3BO_3 in 900 mL water, adjust the pH to 11 with KOH (47% solution) and dilute to 1 L with water. This solution is stable for up to 12 months at 4 °C.
2. *Reducing solution:* Add 2.5 mL of 2-mercaptoethanol to 100 mL buffer solution. This solution is stable for up to 6 months at 4 °C in the dark.
3. *Derivatization reagent:* Dissolve 25 mg of ortho-phthaldialdehyde (OPA) in 2 mL of methanol, mix with 2 mL of reducing solution and dilute to 44 mL with buffer solution. The reagent is stable for up to 7 days at 4 °C in the dark.

[15]Appuhn, A., Joergensen, R. G., Scheller, E. and Wilke, B. 2004. The automated determination of glucosamine, galactosamine, muramic acid and mannosamine in soil and root hydrolysates by HPLC. *J. Plant Nutr. Soil Sci.* 167: 17-21.

4. *Standard stock solutions of the four amino sugars:* Dissolve standards (Sigma Aldrich) in water to a concentration of 1,000 μmol L^{-1} (mannosamine, galactosamine, glucosamine) and 100 μmol L^{-1} (muramic acid), respectively, and store at –18 °C. Prepare standard working solutions by diluting four aliquots of the stock solutions to a concentration range between 210 and 5 μmol L^{-1} (mannosamine, galactosamine, glucosamine) and 21 and 0.5 μmol L^{-1} (muramic acid), respectively. The working standard solutions are stable for over 12 months at −18 °C.

Chromatographic Conditions

Perform chromatographic separations on a column (125 mm length × 4 mm diameter, 5 μm particle size, 12 nm pore size), protected by a security guard cartridge (4 mm length × 2 mm diameter). Place the column in a column oven set at 35 °C. The HPLC system consisting of a gradient pump, an analytical autosampler with in-line split-loop injection and thermostat and a fluorescence detector is set at 445 nm emission and 330 nm excitation wavelength with medium sensitivity. An autosampler designed for automated pre-column derivatization may be used because of the need of a high injection precision and an effective external needle wash for eliminating carryover.

Store the vials with OPA reagent and samples as well as vials for preparation in the autosampler at 15 °C. For derivatization, mix 50 μL OPA reagent and 30 μL sample in a preparation vial and after 120 s reaction time, inject 15 μL of the indole derivates.

The mobile phase consists of two eluents. Eluent *A* is a 97.8/0.7/1.5 (v/v/v) mixture of a water phase, methanol and tetrahydrofuran (THF). Prepare the water phase with 52 mmol sodium citrate and 4 mmol sodium acetate and adjust the pH to 5.3 with HCl. Then, add methanol and THF. Eluent *B* consists of 50% water and 50% methanol (v/v). For degassing and sterilization, both eluents are filtered over 0.2-μm-pore size hydrophilic propylene membrane filters. Deliver the mobile phase at a flow rate of 1.5 mL min^{-1}. Perform the amino sugar separation isocratically, but use a gradient for cleaning the column after every run. Start every run at an eluent *A*/*B* v/v composition of 93/7 for 19 min. Run a linear gradient to reach 80% *B* after 3 min and remaining isocratic for 3 min. A reverse gradient to 93/7 within 3 min is to be followed by 2 min isocratic run. After this, precondition the column for the next sample.

It is important to start the OPA derivatization only at the beginning of every run and not during column preconditioning of the previous run. Otherwise, the equilibration time will not be long enough for a good reproducibility.

20.7. Organic Matter Pools

20.7.1. Active Pools of Carbon

1. Water-soluble Organic Carbon[16]

Water-soluble organic carbon can be extracted from field-moist samples within 24 hours of sampling by shaking 5 g of soil and 10 mL of distilled water for 60 minutes, followed by centrifugation at 10,000 rpm for 30 minutes. Filter the supernatant solution. Take 5 mL of aliquot in a conical flask and treat with 5 mL of 0.07 *N* $K_2Cr_2O_7$, 10 mL of conc. H_2SO_4 and 5 mL of orthophosphoric acid. Mix the sample carefully and digest at 150 °C for 30 minutes using digestion block and cool. Add 1 mL of diphenylamine indicator and titrate against 0.035*N* ferrous ammonium sulphate in 0.4 *M* H_2SO_4. The end point is the appearance of dark green colour. Simultaneously run a blank also.

Calculations

$$\% \text{ Water soluble carbon} = \frac{(B\text{-}S) \times 0.035 \times 0.003 \times 100}{\text{Weight of soil}}$$

Where, B = Blank reading and S = Sample reading

2. Water-soluble Carbohydrates[17]

Air-dry the soil sample and pass through a 1-mm sieve. Hydrolyze 5 g of soil samples for 24 hours on a steam bath with 10 mL of 3*N* H_2SO_4 per g of soil. Place the material in 125 mL Erlenmeyer flasks, covering the mouth by a funnel to minimize evaporation. Maintain the contents at about 85 °C throughout the hydrolysis. Neutralize aliquots of soil hydrolyzates by adding excess of 5 mL of 6*N* NaOH for neutralization. While still hot, pass the hydrolyzates through silica Buchner funnel or sintered glass funnel and wash the residue with 10 mL of hot water per g of soil. Cool the filtrates to room temperature and make to volume for determination. Take 5 mL of aliquot in a test tube, add 10 mL of 0.2% anthrone reagent and shake well. (A freshly prepared 0.2% solution of anthrone is made up in 95% sulphuric acid at least one hour before use). Measure the intensity of green colour at 625 nm using a spectrophotometer. Simultaneously run a blank also for calibrating the instrument. Prepare a series

---

[16]McGill, W. B., Cannon, K. R., Robertson, J. A. and Cook, F. D. 1986. Dynamics of soil microbial biomass and water-soluble organic C in Breton L after 50 years of cropping to two rotations. *Canadian J. Soil Sci.* 66: 1-19.

[17]Brink Jr, R. H., Dubach, P. and Lynch, D. L. 1960. Measurement of carbohydrates in soil hydrolyzates with anthrone. *Soil Sci.* 89: 157-166.

of standard glucose solutions containing 5, 10, 15, 20, 30 mg kg^{-1} from the glucose stock solution (250 mg kg^{-1}) and using the standard curve, obtain the concentration of water- soluble carbohydrates.

Calculations

$$\text{Water-soluble carbohydrates} = \frac{\text{A (ppm)} \times 100}{\text{Weight of soil}}$$

Where, A (ppm) = Concentration of water-soluble carbohydrates read from standard graph

3. Biomass Carbon[18]

Weigh out 12.5 g of oven dried soil into 100 mL beaker. Place the beaker in a 250 mL air- tight plastic container into which add about 5 mL of water. Prepare ethanol-free chloroform immediately before fumigation by passing 100 mL of chloroform through a glass column containing 75 g of basic aluminium oxide. Carry out fumigation with ethanol-free chloroform for 20 hours at 25 °C. After fumigation, remove chloroform by repeated evacuations. After fumigation and removal of chloroform, return the beaker holding the soil to the air tight container together with a scintillation vial holding 5 mL of 0.5 *N* NaOH (fumigated samples) or 0.2 *N* NaOH (non-fumigated samples). Inoculate soil samples with a pinch of fresh soil of respective treatments and incubate the soil for a further period of 10 days at 25 °C. Determine the evolved CO_2 by titrating the alkaline traps with HCl (0.5 *N*) after precipitation of CO_3^{2-} with 50% $BaCl_2$ and using phenolphthalein as indicator. Calculate biomass C from the net amount of (C fumigated – C nonfumigated) CO_2-C evolved, using a K_C factor of 0.45 and multiplying CO_2-C by it.

Calculations

Volume of HCl consumed for sample = A mL

Volume of NaOH taken = 5 mL

Amount of NaOH consumed by evolved CO_2 in the sample = 5-A mL = B mL

Volume of HCl consumed for blank = C mL

Amount of NaOH consumed by evolved CO_2 in the blank = 5-C mL = D mL

1mL of 1 *N* NaOH contains = 22 mg CO_2-C

1mL of 0.5 *N* NaOH contains = 11 mg CO_2-C

[18]Jenkinson, D. S. and Powlson, D. S. 1976. The effects of biocidal treatments on metabolism in soil - V: A method for measuring soil biomass. *Soil Biol. Biochem.* 8: 209 - 213.

$$\text{Total amount of } CO_2\text{-C evolved from the sample (ppm)} = \frac{(B-D)\ mL \times 11 \times 1000}{\text{Weight of soil}}$$

Microbial biomass carbon = (Total amount of CO_2 evolved from fumigated sample - Total amount of CO_2 from non-fumigated sample) × 0.45

4. Biomass Nitrogen[19, 20]

Biomass N can be determined by the fumigation - incubation technique. Conditions for fumigation and incubation are the same as described for determination of biomass C. Calculate the biomass N as the net flush (fumigated - non-fumigated) of N mineralization (measured as ammonium N) over a 10-day incubation after fumigation. Extract ammonium N with 2 *M* KCl and distil an aliquot of 20 mL of the above filtrate with freshly ignited MgO in Bremner distillation apparatus, collect the distillate in 2% boric acid containing mixed indicator and titrate with standard H_2SO_4 (Keeney and Bremner, 1966). Convert the net N flush into biomass N using a K_n factor of 0.57 (Jenkinson, 1988).

Calculations

$$NH_4 \text{ present in the soil (\%)} = \frac{A\ mL \text{ of } H_2SO_4 \text{ consumed} \times 0.00028 \times 100 \times 100}{20 \times \text{Weight of soil}}$$

0.02*N* H_2SO_4 contains = 0.00028 g of N

$$NH_4\text{-N expressed in (mg kg}^{-1}) = \frac{\text{Amount of } NH_4\text{-N (\%)} \times 10^6}{100}$$

Microbial Biomass N (mg kg^{-1}) = (NH_4-N present in fumigated sample – NH_4-N present in non-fumigated sample) × 0.57

[19]Keeney, D. R. and Bremner, J. M. 1966. Comparison and evaluation of laboratory methods of obtaining an index of soil nitrogen availability. *Agron. J.* 58: 498-503.

[20]Jenkinson, D. S. 1988. Determination of microbial biomass carbon and nitrogen in soil. In: *Advances in Nitrogen Cycling in Agricultural Ecosystems*, J. R. Wilson, (Ed.), pp. 368 - 386, CAB International, Wallingford, UK, 1988.

20.7.2. Slow Pools of Carbon[21]

Aggregate Separations

Wet-sieve a 100-g subsample of soil through a series of three sieves to obtain four aggregates fractions: (i) 2000 μm (large macroaggregates), (ii) 250 to 2000 μm (small macroaggregates), (iii) 53 to 250 μm (microaggregates), and (iv) < 53μm (silt plus clay-size particles). Capillary-wet the soil overnight prior to sieving. Bring the capillary-wetted soils consistently up to field capacity plus 5% (w/w), the water content at which maximum aggregate stability is attained. Suspend the capillary-wetted soil samples in water at room temperature on the largest sieve for 5 minutes before sieving. Accomplish aggregate disruption by moving the sieve 3 cm vertically 50 times during a period of 2 minutes. Backwash the material remaining on the sieves into round aluminium pans (23-cm diameter) and dry at 50 °C overnight in a forced-air-oven. Pour the soil plus water that passed through the sieves onto the next finer sieve size and repeat the process. Centrifuge the smallest fraction remaining in the bottom pan (silt plus clay) for 10 minutes at 2500 rpm and backwash the pellet to a round aluminium pan and dry overnight at 50 °C. Weigh the dried aggregate size fractions and store in wide-mouth vials at room temperature.

Prior to chemical analysis, remove plant residue and roots that were larger than 1mm in length by hand with a forceps from subsamples of the aggregate-size fractions. Grind the aggregates in a mortar and pestle and subsample to determine total organic carbon by wet oxidation.

Calculations

$$\%\ \text{Aggregate size fractions of organic C} = \frac{(\text{B-S}) \times 0.035 \times 0.003 \times 100}{\text{Weight of soil}}$$

0.035 = Normality of ferrous ammonium sulphate

20.7.3. Passive Pools of Carbon[22]

1. Major Acid Fractions (Humic substances)

Humic and fulvic acids are estimated by separation after extracting with 0.5 *N* NaOH. Their differential solubility in alkali and acid is adopted as the criterion for separating them into empirical groups.

---

[21]Cambardella, C. A. and Elliott, E. T. 1994. Carbon and nitrogen dynamics of soil organic-matter fractions from cultivated grassland soils. *Soil Sci. Soc. Am. J.* 58: 123-130.

[22]Stevenson, F. J. 1965. Gross chemical fractionation of organic matter. In: *Methods of Soil Analysis, Part 2*. C. A. Black *et al.* (Eds.), pp. 1409 - 1421, American Society of Agronomy, Madison, Wisconsin.

Humic Acid

Take acid-washed (0.1 *N* HCl) 40 g soil sample in a polyethylene centrifuge bottle and add 200 mL of 0.5 *N* NaOH. Shake the mixture for 12 hours in a mechanical shaker and centrifuge at 3000 rpm for 10 minutes. Filter the dark-coloured supernatant liquid and adjust the pH of the filtrate to 1.0 with concentrated HCl. Add additional 200 mL of 0.5 *N* NaOH to the residual soil, shake, centrifuge and filter. Disperse the residue in 200 mL distilled water, centrifuge and add the supernatant liquid to the previous extracts and adjust the pH to 1.0 with concentrated HCl and allow the humic acid to settle down.

The supernatant liquid in the acidified extract is fulvic acid and siphon it off. Transfer the suspension to a polyethylene bottle and centrifuge off the humic acid at 3000 rpm for 10 minutes. Redissolve humic acid in 0.5 *N* NaOH and reprecipitate with concentrated HCl.

Repeat this purification several times. Transfer the supernatant liquid in each case to the original acid filtrate. Wash humic acid with distilled water until free of chloride. Dry the extracted humic acid in a rotary evaporator and grind to a fine powder. Weigh this and report as percentage of humic acid in soil and organic matter.

Calculation of Humic Acid

Weight of humic acid = A mg

$$\text{Humic acid (\%)} = \frac{\text{A mg} \times 100}{40 \times 1000}$$

To express the humic acid content as percentage of organic matter, find the organic C content in the particular soil sample. Multiply the organic C with 1.724 to get the amount of organic matter present in soil.

$$\text{Humic acid (\% of organic matter)} = \frac{100 \times \text{Humic acid (\%)}}{\text{Organic matter (\%)}}$$

Fulvic acid

The acid extract collected in the humic acid preparation is fulvic acid. Take a known aliquot, evaporate and dry. Weigh the residue and report as percentage of fulvic acid on moisture and ash free basis and also as percentage of organic matter.

Calculation of Fulvic Acid

Weight of fulvic acid = A mg

$$\text{Fulvic acid (\%)} = \frac{\text{A mg} \times 100}{40 \times 1000}$$

To express the fulvic acid content as percentage of organic matter, find the organic C content in the particular soil sample. Multiply the organic C with 1.724 to get the amount of organic matter present in soil.

$$\text{Fulvic acid (\% of organic matter)} = \frac{100 \times \text{Fulvic acid}}{\text{Organic matter}}$$

2. Non-humic Substances (Proximate Constituents) of Soil Organic Matter

Soxhlet's extraction procedure can be used to estimate the contents of fats and waxes and resins as ether and alcohol soluble fractions. Do the quantitative estimation of hemicellulose and cellulose after hydrolyzing to convert them into reducing sugars. Use the residue left over after all the above treatment for the determination of lignin. Calculate lignin by analysing the organic carbon content in the H_2SO_4 residue. Calculate protein from the total nitrogen content of the soil residue.

Ether-soluble Fractions (Fats, Waxes, etc.)

Extract 25 g of soil with diethyl ether in Soxhlet's apparatus for 30 hours and gravimetrically determine the ether-soluble fraction after evaporating and drying the extract at 100 °C.

Calculations

Weight of soil sample of before extraction = A g

Weight of soil sample of after extraction = B g

$$\text{Percentage of fats and waxes} = \frac{\text{(A-B)} \times 100}{\text{Weight of soil}}$$

Alcohol-soluble Fractions (Resins)

Treat the soil after ether extraction with 95% ethanol for 2 hours on a hot water bath under reflux and filter through Whatman No. 50 filter paper. Filter after repeated washings with alcohol. Estimate the quantity of resins present gravimetrically by evaporating the alcohol and drying the residue at 100 °C.

Calculations

Weight of soil sample of before extraction = A g

Weight of soil sample of after extraction = B g

$$\text{Percentage of resins} = \frac{(A\text{-}B) \times 100}{\text{Weight of soil}}$$

Hot Water-soluble Fractions (Carbohydrates / Free Sugars)

After alcohol extraction, boil the soil residue under reflux with hot water for 2 hours and filter. Wash the residue repeatedly with hot water for collecting the filtrate and make up its volume to 1000 mL. Evaporate 100 mL of this solution on a steam bath and dry the residue in an oven at 100 °C for 30 minutes. Keep the dried residue later in a muffle furnace at 550 °C and weigh, and it constitutes the water-soluble polysaccharides.

Calculations

$$\text{Percentage of free sugars} = \frac{\text{Weight of residue} \times 1000 \times 100}{100 \times \text{Weight of soil}}$$

Hemicellulose

Dry the soil residue from the hot water extract in an oven at 100 °C and extract the soil with 300 mL of 2% HCl solution for 5 hours under reflux and filter using a Buchner funnel. Wash the soil residue thoroughly with HCl solution and make up the volume of the filtrate to 500 mL. Take 10 mL aliquots of this filtrate in a 250 mL Erlenmeyer flask and add 2 drops of bromocresol purple indicator. Add 2.5% NaOH solution to the Erlenmeyer flask until the colour turns purple and allow to stand for few hours. Filter the mixture, wash with distilled water and make up to 100 mL. Add 10 mL of Fehling's solution to the above mixture and boil for few minutes. Estimate the reducing sugars present in the above mixture by titrimetry by adding 3 to 5 drops of methylene blue indicator and using 0.2% standard glucose solution as titrant. From the quantity of the reducing sugars present, calculate the hemicellulose content of the soil.

Calculations

$$\text{Percentage of Hemicellulose} = \frac{\text{mL of std. glucose consumed} \times 0.2 \times 0.9 \times 100 \times 500 \times 100}{1000 \times 10 \times \text{Weight of soil}}$$

1mL of standard 0.1N contains = 0.2 mg of hemicellulose

0.9 is a factor which takes into account the loss of water accompanying the polymerisation of monosaccharides.

Cellulose

Dry the soil residue obtained after HCl extraction in an oven and mix 2 g of the soil with 25 mL of 30 N sulphuric acid solution and allow to stand for 2.5 hours at 12 °C to 14 °C. After that, add 875 mL of water to it and boil under reflux for 5 hours. Use the filtrate obtained for estimating the reducing sugars, following the same procedure as that of hemicellulosc.

Calculations

$$\text{Percentage of Cellulose} = \frac{\text{mL of std. glucose consumed} \times 0.2 \times 0.9 \times 100 \times 1000 \times 100}{1000 \times 10 \times \text{Weight of soil}}$$

1mL of standard 0.1N contains = 0.2 mg of cellulose

0.9 is a factor which takes into account the loss of water accompanying the polymerisation of monosaccharides.

Lignin

Dry the soil that remained after cellulose estimation to a constant weight and take a portion of this finely powdered soil to estimate the organic carbon following the method of Walkley and Black (1934). Calculate the total organic matter present in the residue by multiplying the organic carbon content with 1.724.

Calculations

$$\text{Organic C \%} = \frac{(\text{B-S}) \times 0.035 \times 0.003 \times 100}{\text{Weight of soil}}$$

Lignin (%) = Organic C (%) × 1.724

Protein

Analyze the same residue for its total nitrogen content and calculate the protein content by multiplying the total nitrogen content with the factor 6.25.

Calculations

$$\text{Total N (\%)} = \frac{\text{mL of } 0.02\ N\ H_2SO_4 \text{ consumed} \times 0.00028 \times 100}{\text{Weight of soil}}$$

Protein (%) = Total N (%) × 6.25

Further Reading

Carter, M. R. and Stewart, B. A. (Eds.) 2019. *Structure and Organic Matter Storage in Agricultural Soils.* CRC Press, USA.

Garcia, C., Nannipieri, P. and Hernandez, T. 2018. *The Future of Soil Carbon.* Academic Press, New York.

IAEA. 2017. *Use of Carbon Isotopic Tracers in Investigating Soil Carbon Sequestration and Stabilization in Agroecosystems.* Tecdoc No. 1823, IAEA, Vienna.

IAEA. 2019. *Use of Laser Carbon Dioxide Carbon Isotope Analysers in Agriculture.* Tecdoc No. 1866, IAEA, Vienna.

Kononova, M. M. 2013. *Soil Organic Matter: Its Nature, its Role in Soil Formation and in Soil Fertility.* Elsevier Science, Amsterdam, The Netherlands.

Magdoff, F. and Weil, R. R. 2019. *Soil Organic Matter in Sustainable Agriculture.* CRC Press, USA.

Vaughan, D. and Malcolm, R.E. (Eds.) 2013. *Soil Organic Matter and Biological Activity.* Springer Science + Business Media, New York.

0

Colour Plates

Subject Index